# THE SECRETS OF

# TRIANGLES

# THE SECRETS OF

# TRIANGLES

## ~ A MATHEMATICAL JOURNEY ~

### ALFRED S. POSAMENTIER
### AND INGMAR LEHMANN

59 John Glenn Drive
Amherst, New York  14228–2119

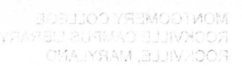

Published 2012 by Prometheus Books

Cover image © Media Bakery/Glenn Mitsui
Jacket design by Jacqueline Nasso Cooke

Inquiries should be addressed to
Prometheus Books
59 John Glenn Drive
Amherst, New York 14228–2119
VOICE: 716–691–0133
FAX: 716–691–0137
WWW.PROMETHEUSBOOKS.COM

16 15 14 13 12    5 4 3 2 1

Library of Congress Cataloging-in-Publication Data

Posamentier, Alfred S.
    The secrets of triangles : a mathematical journey / by Alfred S. Posamentier and Ingmar Lehmann.
        p. cm.
    Includes bibliographical references and index.
    ISBN 978–1–61614–587–3 (cloth : alk. paper)
    ISBN 978–1–61614–588–0 (ebook)
    1. Trigonometry. 2. Triangle. I. Lehmann, Ingmar. II. Title.

QA531.P67 2012
516'.154—dc23

2012013635

Printed in the United States of America on acid-free paper

To Barbara for her support, patience, and inspiration.

To my children and grandchildren, whose future is unbounded: David, Lauren, Lisa, Danny, Max, Sam, and Jack.

And in memory of my beloved parents, Alice and Ernest, who never lost faith in me.

Alfred S. Posamentier

To my wife and life partner, Sabine, without whose support and patience my work on this book would not have been possible.

And to my children and grandchildren: Maren, Claudia, Simon, and Miriam.

Ingmar Lehmann

# CONTENTS

# ACKNOWLEDGMENTS

The authors wish to extend sincere thanks for proofreading and useful suggestions to Dr. Michael Engber, professor emeritus at the City College of the City University of New York; Dr. Manfred Kronfeller, professor of mathematics at Vienna University of Technology, Austria; Dr. Bernd Thaller, professor of mathematics at Karl Franzens University–Graz, Austria; and Dr. Peter Schöpf, professor of mathematics at Karl Franzens University–Graz, Austria. We are especially grateful to Dr. Robert A. Chaffer, professor of mathematics at Central Michigan University for contributing chapter 10 on fractals. We also thank Catherine Roberts-Abel for managing the production of this book and Jade Zora Ballard for her meticulous editing.

# PREFACE

The triangle is one of the basic structures of geometry. We see it in many patterns, and we find that many geometric structures can be best analyzed by partitioning them down into triangles. Yet triangles provide one of the richest examples of geometric phenomena that allow us to admire the beauty of geometry. This is what we hope to achieve in this book, using nothing more than the geometric concepts presented in high-school geometry courses.

When we hear the word *triangle* we tend to recall some of the special triangles that we encountered many times in the past, beginning in our earliest days of schooling with such triangles as the equilateral triangle, the isosceles triangle, and the scalene triangle, which were classified by their side lengths. At times we considered triangles classified by their angles, such as the right triangle, the acute triangle, and the obtuse triangle. An enlightened teacher probably alerted us to the fact that these triangle descriptions were words that came from our everyday English language outside of mathematics. *Equilateral* means equal-sided, as the word *lateral* refers to side. *Isosceles* comes from the Greek word *iso*, which means equal; and the Greek word *isoskeles* means equal-legged. The term *scalene* stems from the Latin *scalenus* or the Greek *skalenos*, meaning unequal. A right triangle is one that is erect, coming from the German word *recht*, which comes from the Latin *rectus*, meaning upright, as in perpendicular to a horizontal line.

When we speak of an acute pain, we refer to a sharp pain, hence the word *acute* means sharp. And so an acute triangle is one that has all sharp angles. A dull person is often referred to as being *obtuse*, or not sharp and clear. Consequently, an obtuse triangle is one that has a dull angle.

This classification of triangles is essentially what many people recall about triangles from school days. Some may even recall that there were certain intriguing relationships that occurred in all triangles, such as that the three altitudes (the line segments from a vertex drawn perpendicular to the opposite side) are always concurrent (i.e., intersect at a common point), as are the three angle bisectors (the lines that divide an angle into two equal parts) and the three medians (lines that join a vertex with the midpoint of the opposite side). There are boundless other beautiful properties of triangles—many of which are truly amazing—that we shall explore in this book. Concurrencies arise when one would least expect them. For example, suppose we inscribe a circle in any randomly drawn triangle and then join the three points of tangency to each to the opposite vertices, then we find that these three lines are concurrent. And that is true for *all* triangles! This was first discovered by the French mathematician Joseph-Diaz Gergonne (1771–1859). We will expand on this surprising relationship in the pages that follow.

Another truly amazing triangle relationship that will be among the many aspects of the triangle we will explore is called *Morley's theorem*, named after Frank Morley (1860–1937), who was the father of the famous American author Christopher Morley. In 1899 he discovered that if one trisects each of the angles of a triangle—regardless of the shape or size of the triangle—the adjacent angle trisectors will *always* meet at three points forming an equilateral triangle. We will explore this astonishing property (and other related relationships) and even *prove* that it is really true for all triangles!

A remarkable relationship between the interior lines of a triangle (such as the ones we described above) and the sides of the triangle was discovered by the Italian mathematician Giovanni Ceva (1647–1734) in 1678. This theorem makes proving concurrency almost trivial, where traditional proofs—not using Ceva's theorem—would be quite cumbersome. We will also consider the analogue of this lovely relationship—discovered by Menelaus of Alexandria (70–130 CE)—to easily establish if three given points lie on the same straight line.

Besides exploring the multitude of surprising relationships connected with triangles—both special and general triangles—we will also show how and when triangles can be constructed using a straightedge and compasses. This is perhaps the one opportunity in geometry where genuine problem-solving techniques are best and most simply exhibited. It is today a rather neglected aspect of geometric explorations, yet one that will appeal to all by the cleverness of the approaches used in doing these relatively elementary constructions—the simplest of which is likely the one most readers will recall from their high-school geometry course, the construction of a triangle, given the lengths of its three sides. Yet, we can also—and very cleverly—construct a triangle given only the lengths of its three altitudes. Fun with such triangle constructions will sharpen problem-solving skills.

To make our book reader friendly, we will use a very simple language— one that was used in high-school geometry books in past years. We will avoid using some of the more modern (and more precise) nomenclature, again, to make it easier to read. We will call a line, $AB$; a line segment, $AB$; a ray, $AB$; and the measure of a line segment, $AB$, all with the designation $AB$ to make the reading a bit less cumbersome. We also do not expect the reader to be familiar with the accepted designations for various triangle parts, such as the center of the inscribed circle of a triangle usually being designated by the letter $I$, or the centroid usually being designated by the letter $G$. We use convenient letters for each diagram that we feel will be reader friendly. To further make our discussions clear to the reader, we provide diagrams for all these discussions—something not necessarily common to all geometry books. We are truly focusing on concept clarity!

As we are about to embark on a journey of exploration of triangle properties that are possessed by special triangles, such as the right triangle (yes, also including the famous Pythagorean theorem), the equilateral triangle, the isosceles triangle, and, of course, the general triangle. We will construct triangles and then we will admire the brilliance of those who discovered the many hidden treasures of geometry. So join us now on this bountiful exploration of all aspects of one of the most common, yet mighty, of geometric figures: the triangle!

# CHAPTER 1

# INTRODUCTION TO THE TRIANGLE

Arithmetic! Algebra! Geometry!
Grandiose trinity, brilliant triangle!
Who has not known you, is a poor wretch!
. . . But who knows you and appreciates you,
desires no further goods of the earth.
—*The Songs of Maldoror* II, 10[1]

The word *triangle* is used in a variety of contexts. For example, there is the *Bermuda Triangle*, which refers to the area of a triangle determined by three points: one at Miami, Florida; another at San Juan, Puerto Rico; and a third at Bermuda. It is believed that this triangular surface has had an inordinate number of ship and aircraft mishaps. There is also another well-known triangular region called the *Summer Triangle*: three stars that determine a triangle. The summer triangle consists of the stars known as Deneb, Altair, and bluish Vega. The American essayist Henry David Thoreau (1817–1862) has been often quoted with the following: "The stars are the apexes of what triangles!"

Then there is the *culinary triangle*, a concept described by anthropologist Claude Lévi-Strauss (1908–2009) involving three types of cooking; these are boiling, roasting, and smoking, usually done to meat. Here the triangle is determined by the three sides or angles, depending on how it is used. Then there is the *social triangle* as described by the French writer Honoré de Balzac (1799–1850). The three points of the social triangle are skill, knowledge, and capital. Another triangle determined by the three sides is the musical instrument *the triangle*. What we then have is a variety of ways that we can define a triangle geometrically: either a polygon of three sides, or three noncollinear points, or the area within the region determined by the previous two definitions.

The triangle is the basic geometrical figure that allows us to best study geometrical shapes. A quadrilateral can be partitioned into two triangles, a pentagon into three triangles, a hexagon into four triangles, and so on. (See figures 1-1a, 1-1b, and 1-1c.) These partitions allow us to study the characteristics of these figures. And so it is with Euclidean geometry—the triangle is one of the very basic parts on which most other figures depend.

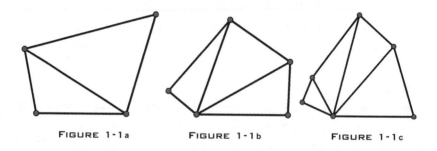

FIGURE 1-1a          FIGURE 1-1b          FIGURE 1-1c

Yet before we embark on our journey investigating triangles and their many related line segments and angles, we ought to determine what it takes for a triangle to exist. Suppose you have three rods and the sum of the lengths of two of them is shorter than the length of the third rod, then you will see that you cannot form a triangle with the three rods. (See figure 1-2.)

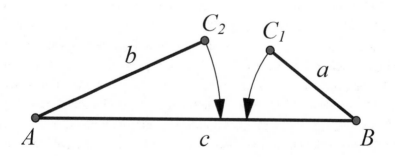

FIGURE 1-2

We can generalize this by saying that in order for a triangle to exist, the sum of the lengths of any two sides must be greater than the third side.

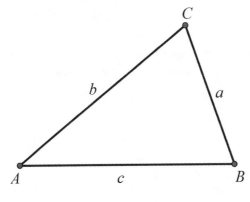

In figure 1-3, for triangle *ABC* we have the following inequalities:

$$a + b > c$$
$$a + c > b$$
$$b + c > a$$

## TRIANGLE CONGRUENCE

Let us now review the various relationships that can connect two triangles. First there is the *congruence* of two triangles (using the symbol ≅), which describes two triangles with the exact same size and shape so that they can be placed to perfectly coincide. In other words, the corresponding sides and angles of the two triangles are equal. To show that two triangles are congruent, we do not need to determine that all the corresponding sides and angles are equal. Rather we can establish the congruence of two triangles simply by showing that any one of the following is true:

- The three sides of one triangle ($\triangle ABC$) are equal to the three corresponding sides of the other triangle ($\triangle DEF$).
- Two right triangles can be shown to be congruent if the hypotenuse and a leg of one triangle are equal to the corresponding sides of the second triangle.
- Two sides and the angle between them of one triangle ($\triangle ABC$) are equal to corresponding parts of the other triangle ($\triangle DEF$). (See figure 1-4.)

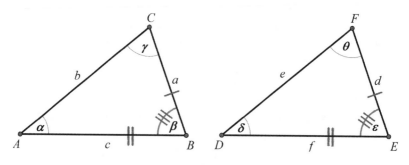

FIGURE 1-4

- Two angles and one side of one triangle ($\triangle ABC$) are equal to the corresponding parts of the other triangle ($\triangle DEF$).

We indicate this congruence symbolically as $\triangle ABC \cong \triangle DEF$.

Another relationship between two triangles is *similarity* (represented by the symbol ~), which tells us that the two triangles have the same shape but not necessarily the same size, that is, that the corresponding angles of the two triangles are equal. Similarity between two triangles can be established by showing that:

- Two angles of one triangle ($\triangle ABC$) are equal to two angles of the other triangle ($\triangle DEF$) as shown in figure 1-5.

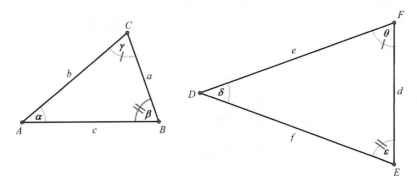

FIGURE 1-5

- The three sides of one triangle ($\triangle ABC$) are proportional to the three sides of the other triangle ($\triangle DEF$).
- Two sides of one triangle ($\triangle ABC$) are proportional to two sides of the other triangle ($\triangle DEF$) and the angles between these two sides of each triangle are congruent.

Symbolically we write this as $\triangle ABC \sim \triangle DEF$.

Two triangles can also be related by their position in the plane. For example, consider two triangles, $\triangle ABC$ and $\triangle A'B'C'$ ( of possibly different shapes), whose corresponding sides (extended) meet in three collinear points $X$, $Y$, and $Z$ (i.e., points that lie on the same straight line):

sides $AC$ and $A'C'$ meet at point $X$,
sides $BC$ and $B'C'$ meet at point $Y$, and
sides $AB$ and $A'B'$ meet at point $Z$.

Then the lines joining the corresponding vertices ($AA'$, $BB'$, and $CC'$) are concurrent (in point $P$), as shown in figure 1-6. This famous two-triangle relationship was first discovered by the French mathematician and engineer Gérard Desargues (1591–1661) and today bears his name. The converse of this relationship is also true. Namely, if two triangles are so placed that the lines joining their corresponding vertices are concurrent (in figure 1-6, point $P$ is that point of concurrency), then the extensions of their corresponding sides will meet in three collinear points (points $X$, $Y$, and $Z$).

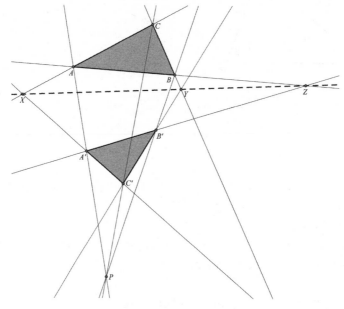

**FIGURE 1-6**

## THE EQUILATERAL TRIANGLE

There are also triangles that have special relationships within themselves. Perhaps the most common is the *equilateral triangle*, which is one that has all sides equal and all angles equal. Not only that, but all of its angle bisectors, altitudes, and medians are equal to each other. A lesser-known property of the equilateral triangle is seen by taking any point, $P$, in an equilateral triangle (figure 1-7) and drawing the perpendicular segments to each of its three sides. The sum of the distances from this randomly chosen point to the three sides, $PQ + PR + PS$, is always the same. That sum is equal to the altitude of the equilateral triangle. This is shown in figure 1-7, where the altitude is $CD$. This relationship, often called *Viviani's theorem*, is attributed to the Italian mathematician Vincenzo Viviani (1622–1703), who, incidentally, was a student of the famous Italian scientist and philosopher Galileo Galilei (1564–1642).

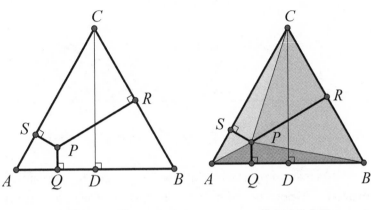

FIGURE 1-7                    FIGURE 1-8

This surprising property can be proven by using the formula for the area of a triangle (i.e., the area of a triangle is one-half the product of the base and the altitude drawn to that base). We begin with equilateral triangle $ABC$, where $PR \perp BC$, $PQ \perp AB$, $PS \perp AC$, and $CD \perp AB$. We then draw $PA$, $PB$, and $PC$, as shown in figure 1-8.

The $Area\ \triangle ABC = Area \triangle APB + Area \triangle BPC + Area \triangle CPA$
$= \frac{1}{2}\ AB \cdot PQ + \frac{1}{2}\ BC \cdot PR + \frac{1}{2}\ AC \cdot PS$.

Since $AB = BC = AC$, the $Area \triangle ABC = \frac{1}{2}\ AB \cdot (PQ + PR + PS)$.

However, the $Area \triangle ABC = \frac{1}{2}\ AB \cdot CD$. Therefore, $PQ + PR + PS = CD$, which is then a constant that we sought to prove true for the given triangle.

There are many other relationships special to the equilateral triangle beyond the basic ones we just mentioned. The surprising properties of the equilateral triangle will be presented a bit later. In the meantime, we shall survey some other special triangles. The *isosceles triangle* is one that has at least two sides of the same length. Its base angles are always equal to each other. We will be revisiting the isosceles triangle in our discussions throughout the book.

## THE RIGHT TRIANGLE

The *right triangle* is so named because it has one right angle, as shown in figure 1-9, where $\angle ACB = 90°$.

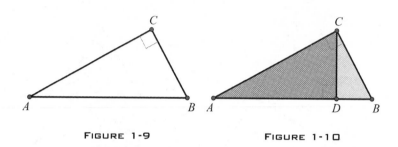

FIGURE 1-9          FIGURE 1-10

It, too, has many properties within itself. For example, when an altitude is drawn to the hypotenuse of the right triangle, the triangle is partitioned into three similar triangles. In figure 1-10, the three similar triangles are $\triangle ABC \sim \triangle ACD \sim \triangle BCD$.

If we look at these three triangles in pairs, we can establish a rather familiar relationship. Follow along!

We will begin with $\triangle ABC \sim \triangle ACD$. From this similarity we get the following proportion of their side lengths: $\frac{AB}{AC} = \frac{AC}{AD}$. This gives us $AC^2 = AB \cdot AD$. From the similarity $\triangle ABC \sim \triangle BCD$, we get $\frac{AB}{BC} = \frac{BC}{BD}$, or $BC^2 = AB \cdot BD$. When we add these two equations, the following results: $AC^2 + BC^2 = AB \cdot (AD + BD) = AB^2$. When we express this verbally, we have the following statement: "The sum of the squares of the legs of a right triangle equals the square of the hypotenuse."

This should remind us of perhaps the most famous theorem in geometry, the *Pythagorean theorem*. If we replace "of" with "on" in this statement, we have, referring to the areas of the squares, "the sum of the squares *on* the legs of a right triangle is equal to the square *on* the hypotenuse.[2]

This can then be shown geometrically as in figure 1-11, namely, the sum of the areas of the two smaller squares (those on the legs of the right triangle) is equal to the area of the larger square—the one on the hypotenuse.

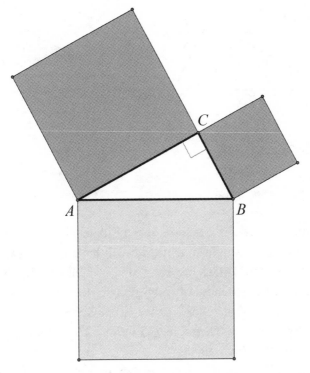

FIGURE 1-11

We know that any three noncollinear points determine a unique triangle as well as a unique circle. However, when the triangle is a right triangle, then the circumscribed circle's diameter is the hypotenuse of the triangle, as shown in figure 1-12, where hypotenuse $AB$ is the diameter of the circumscribed circle $c$ with the midpoint $O$.

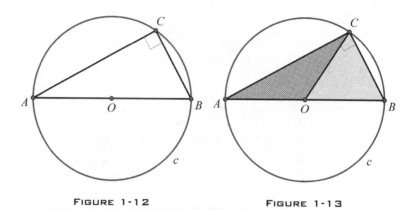

FIGURE 1-12                    FIGURE 1-13

From this property, we can show very easily that if we draw the median to the hypotenuse of a right triangle, we will have formed two isosceles triangles, $\triangle AOC$ and $\triangle BOC$. We show this in figure 1-13, where $CO$ is the median to the hypotenuse of right triangle $ABC$, but in this case, the median—which is also the radius of the circumscribed circle—is half the length of the hypotenuse. Therefore, $CO = BO = AO$. Consequently, $\triangle AOC$ and $\triangle BOC$ are isosceles triangles.

As we mentioned earlier, just as the right triangle is categorized by one of its angles—the angle of 90-degree measure—other triangles can also be categorized by a triangle's angle measures. When a triangle has an angle greater than 90°, which is called an obtuse[3] angle, the triangle is called an *obtuse triangle*. When all of a triangle's angle measures are each less than 90° (i.e., acute[4] angles), then we call the triangle an *acute triangle*.

An extension of the Pythagorean theorem allows us to establish relationships among the sides of a triangle that will help us to determine if a nonright triangle is acute or obtuse.

For a triangle whose sides have lengths $a, b,$ and $c$, if $a^2 + b^2 > c^2$, then the angle between the sides of length $a$ and $b$ is acute (see figure 1-14) and the triangle is an acute triangle.

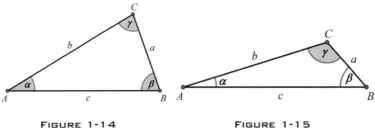

FIGURE 1-14                    FIGURE 1-15

On the other hand, if a triangle's sides have lengths $a$, $b$, and $c$, and if $a^2 + b^2 < c^2$, then the angle between the sides of length $a$ and $b$ is obtuse (see figure 1-15), and the triangle is then an obtuse triangle.

Moreover, for an obtuse triangle, such as $\triangle ABC$, shown in figure 1-16, we have the following relationship, which derives directly from the Pythagorean theorem: $c^2 = a^2 + b^2 + 2ax$. In other words, this would make $c^2$ greater than $a^2 + b^2$, which we stated before.

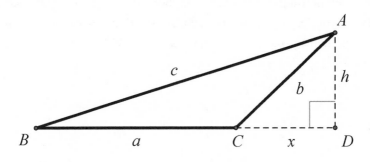

FIGURE 1-16

Now, to show why the equation $c^2 = a^2 + b^2 + 2ax$ is true. Using figure 1-16, and applying the Pythagorean theorem (first to triangle $ABD$): $c^2 = (a + x)^2 + h^2 = a^2 + x^2 + 2ax + h^2 = a^2 + (x^2 + h^2) + 2ax$. However, applying the Pythagorean theorem to triangle $ADC$, we get $b^2 = x^2 + h^2$. Therefore, $c^2 = a^2 + b^2 + 2ax$, which is what we set out to show above.

For an acute triangle, shown in figure 1-17, we have $c^2 = a^2 + b^2 - 2ax$. This can be verified in a manner similar to the method used above, and would justify that $c^2$ is less than $a^2 + b^2$, which we also stated before.

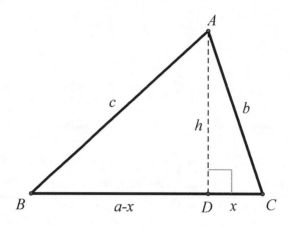

**FIGURE 1-17**

The Pythagorean theorem allows us to arrive at lots of interesting triangle relationships. For example, there is one that is attributed to Apollonius of Perga (ca. 262–ca. 190 BCE), which states that for triangle *ABC*, with median *AD*, we can show that $AB^2 + AC^2 = 2\,(BD^2 + AD^2)$. (See figure 1-18.[5])

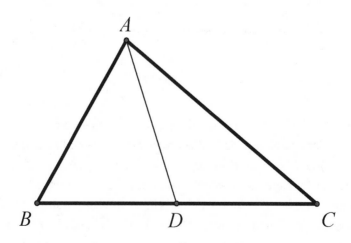

**FIGURE 1-18**

## TRIANGLE AREA

We can determine the area of a triangle in a number of ways depending on the information given about the triangle. If we are given the length of one side of the triangle and the length of the altitude drawn to that side, then we can use our familiar formula for the area: one-half the product of the base and its height. Symbolically that is written as $Area = \frac{1}{2}bh$. (See figure 1-19.)

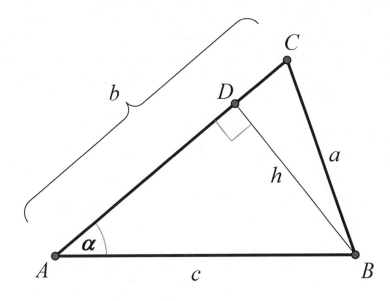

FIGURE 1-19

If we are given the measure of one angle, for example, $\angle A = \alpha$, of a triangle $ABC$ and the lengths of the two sides forming that angle, $b$ and $c$, then we have the following additional formula for the area of triangle $ABC$. Symbolically that is written as $Area\ \triangle ABC = \frac{1}{2}bc \cdot \sin \angle A = \frac{1}{2}bc \cdot \sin \alpha$.

It is also possible to establish the area of a triangle given the lengths $a$, $b$, and $c$, of the three sides of triangle $ABC$ using Heron's formula[6] $Area\ \triangle ABC = \sqrt{s(s-a)(s-b)(s-c)}$, where $s = \frac{1}{2}(a + b + c)$ is the *semi-perimeter* of the triangle $ABC$.

We will be exploring the area of triangles and other related areas in chapter 7.

## TRIGONOMETRY AND THE TRIANGLE

The Pythagorean theorem is actually the basis for all of trigonometry,[7] therefore, of the over four hundred proofs of the Pythagorean theorem that exist today, none uses trigonometry — or else we would have circular reasoning. (Remember, you cannot prove a theorem using a relationship that depends on that theorem!) Yet, with the advent of trigonometry, we have some very useful relationships surrounding the triangle. Each named after one of the three basic trigonometric functions: *sine*, *cosine*, and *tangent*.

Let us first review these basic functions as they apply to a right triangle, and then provide their application to the general triangle. For the right triangle, $\triangle ABC$ (figure 1-20), we have three trigonometric functions defined for $\angle A$ as follows:

$$\sin \angle A = \frac{a}{c}$$

$$\cos \angle A = \frac{b}{c}$$

$$\tan \angle A = \frac{a}{b}$$

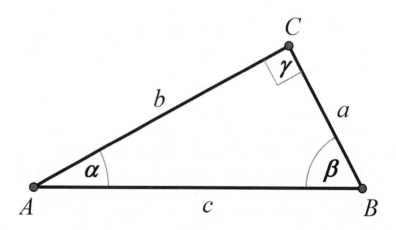

FIGURE 1-20

Extending these trigonometric functions to the general triangle we get the following relationships, known as the *law of sines*:

$$\frac{a}{\sin \angle A} = \frac{b}{\sin \angle B} = \frac{c}{\sin \angle C}$$

It is interesting to see how easily this relationship evolves from the basic above-mentioned definitions of the sine function. To begin, we will consider triangle *ABC*, with altitude *CD* (= $h_c$) to side *AB* (= *c*), which partitions the triangle into two right triangles, $\triangle ACD$ and $\triangle BCD$. (See figure 1-21.)

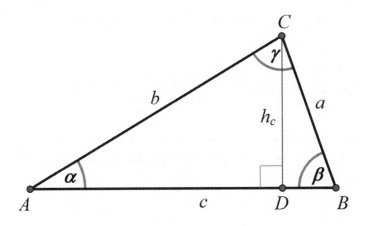

FIGURE 1-21

In the right triangles $\triangle ACD$ and $\triangle BCD$, we can apply the sine function as follows:

$$\sin \alpha = \sin \angle A = \frac{CD}{AC} = \frac{h_c}{b}, \text{ and}$$

$$\sin \beta = \sin \angle B = \frac{CD}{BC} = \frac{h_c}{a}.$$

Therefore, $h_c = b \cdot \sin \angle A = a \cdot \sin \angle B$, or $\frac{a}{\sin \angle A} = \frac{b}{\sin \angle B}$. (We can also write this as $\frac{a}{\sin \alpha} = \frac{b}{\sin \beta}$.)

Had we originally selected the altitude $h_a$ or $h_b$ instead of $h_c$, we would then be able to extend this to get the following relationship we set out to justify, namely, the law of sines: $\frac{a}{\sin \angle A} = \frac{b}{\sin \angle B} = \frac{c}{\sin \angle C}$. This can also be written as $\frac{a}{\sin \alpha} = \frac{b}{\sin \beta} = \frac{c}{\sin \gamma}$.

For the cosine function, we can develop a *law of cosines*:

$$a^2 = b^2 + c^2 - 2bc \cdot \cos \angle A,^{8}$$

or written another way as

$$\cos \angle A = \frac{b^2 + c^2 - a^2}{2bc}, \text{ or } \cos \alpha = \frac{b^2 + c^2 - a^2}{2bc}.$$

For the tangent function, we have the *law of tangents*:

$$\frac{a-b}{a+b} = \frac{\tan\frac{1}{2}(\angle A - \angle B)}{\tan\frac{1}{2}(\angle A + \angle B)}, \text{ or it may be written as } \frac{a-b}{a+b} = \frac{\tan\frac{\alpha-\beta}{2}}{\tan\frac{\alpha+\beta}{2}}.$$

Although the justifications for these latter two relationships is a bit more complex than the one we used for the law of sines, they can be found in most high-school textbooks.

## GOLDEN TRIANGLES

As we conclude our introduction to the triangle, we will present the reader with one of the most beautiful triangles: the golden triangle, one that carries the golden ratio[9] throughout. This is simply an isosceles triangle with a 36° vertex angle and two 72° base angles (figure 1-22).

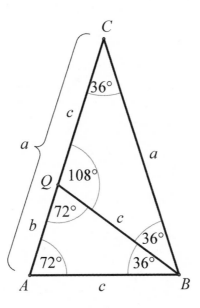

FIGURE 1-22

Triangle $ABC$ is called a *golden triangle*. When we draw the bisector of $\angle ABC$, we have created two similar triangles, $\triangle ABC \sim \triangle AQB$, with the angles as shown in figure 1-22. This enables us to set up the following proportion: $\frac{AC}{AB} = \frac{AB}{AQ}$, or $\frac{b+c}{c} = \frac{c}{b}$.

This is known as the *golden ratio* (often referred to as $\phi$) as the point $Q$ partitions the side $AC$ into the *golden section*. As you can see here: $\frac{b+c}{c} = \frac{c}{b} = \frac{\sqrt{5}+1}{2} = \phi$, or put another way, $\phi{:}1$, or $\frac{a}{c} = \frac{\phi}{1} = \phi$.

In figure 1-22 we have three golden triangles:[10] two with an acute vertex angle (the isosceles triangles $ABC$ and $ABQ$), where the ratio of the lengths of the side to the base is $\phi{:}1$, or $\frac{a}{c} = \frac{\phi}{1}$; and one with an obtuse vertex angle (isosceles triangle $BCQ$), where the ratio of the lengths of the side to the base is $1{:}\phi$, or $\frac{c}{a} = \frac{1}{\phi}$.

Throughout the following chapters we will explore all sorts of triangles and their many surprising relationships. Of course, we will explore how other geometric parts can interact with triangles to exhibit some truly amazing and, yes, *surprising* relationships. Join us now on this awe-inspiring adventure!

# CHAPTER 2

# CONCURRENCIES OF A TRIANGLE

T he lines and points of a triangle combined with the circles that either enclose them or are enclosed by them are the key to many secrets embedded in triangles. In this chapter we will provide some of the most fascinating and surprising relationships of these triangle-related parts. Many of these relationships were not known to Thales (ca. 624–ca. 546 BCE), Pythagoras (ca. 570–ca. 510 BCE), Euclid (ca. 300 BCE), and Archimedes (ca. 287–ca. 212 BCE), our forefathers in this visual part of mathematics. As a matter of fact, Euclid in his famous book, *Elements*, mentions the centers of the inscribed and circumscribed circles of a triangle and with them the concurrent lines that determine these center points, namely, the angle bisectors of the three angles of the triangle and the perpendicular bisectors of the sides of the triangle, respectively. It was not until Archimedes's contributions that there was mention of the altitudes and the medians of a triangle. Till about the end of the eighteenth century, only five significant points relating to the triangle were known. However, not until the nineteenth century did geometry begin to blossom through the work of some famous mathematicians including Joseph-Diaz Gergonne (1771–1859), Jakob Steiner (1796–1863), Karl Wilhelm Feuerbach (1800–1834), Christian Heinrich von Nagel (1803–1882), Joseph Jean Baptiste Neuberg (1840–1926), and many others. More on these mathematicians later! Meanwhile, today there are more than 3,600 noteworthy points related to the triangle. There are many more discoveries about triangle parts that constantly appear in professional journals under "new discoveries." We hope that the reader's exposure to this chapter will entice further exploration and perhaps even some new discoveries.

## INTRODUCTION TO THE ALTITUDES OF A TRIANGLE

When one thinks of lines related to a triangle—other than its sides—one generally thinks of the altitudes,[1] angle bisectors,[2] and medians.[3]

Aside from the properties that define these special line segments, each of these groups of three are concurrent line segments and each point of concurrency is a significant point in the triangle.

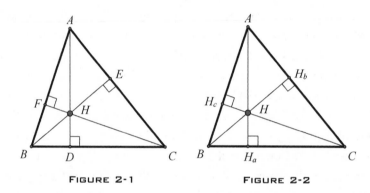

FIGURE 2-1                    FIGURE 2-2

In figure 2-1, we have the three altitudes of triangle *ABC*. Beside the fact that they are perpendicular to the sides of the triangle, we can take note that the point *H* of concurrency—called the *orthocenter* of the triangle—partitions each of the altitudes into two line segments, the product of which is the same for each of the three altitudes. That is, for triangle *ABC* in figure 2-1: $AH \cdot DH = CH \cdot FH = BH \cdot EH$. This evolves from the similarity of triangles in this configuration. That is, since

$$\triangle AFH \sim \triangle CDH, \frac{AH}{FH} = \frac{CH}{DH} .$$ This gives us $AH \cdot DH = CH \cdot FH$.

Similarly,

$$\triangle AEH \sim \triangle BDH,$$ therefore, $\frac{AH}{EH} = \frac{BH}{DH}$ , and then $AH \cdot DH = BH \cdot EH$.

In figure 2-2, where we renamed the feet of the altitudes with subscripts, we can express this relationship as

$$AH \cdot HH_a = BH \cdot HH_b = CH \cdot HH_c.$$

This leads to a rather unusual relationship in the triangle. If we form rectangles from the two parts of each of the cut altitudes, we find that they determine three equal-area rectangles, as shown in figure 2-3.

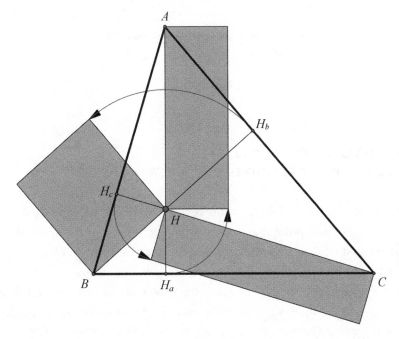

FIGURE 2-3

In figure 2-4, we once again show a triangle with altitudes drawn, and their "feet" (that is, the point at which the altitude intersects with the base) marked as $H_a$, $H_b$, and $H_c$, and the segments along the sides as $a_1 = BH_a$, $a_2 = H_aC$, $b_1 = CH_b$, $b_2 = H_bA$, $c_1 = AH_c$, and $c_2 = H_cB$. This allows us to state easily a relationship discovered in 1828 by the Swiss mathematician Jakob Steiner (1796–1863),[4] namely, that $a_1^2 + b_1^2 + c_1^2 = a_2^2 + b_2^2 + c_2^2$.

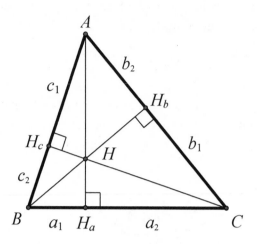

FIGURE 2-4

## INTRODUCTION TO THE
## ANGLE BISECTORS OF A TRIANGLE

The angle bisectors of a triangle, besides bisecting the angles of the triangle, meet at a point that is equidistant from each of the sides of the triangle, and is, therefore, the center of the circle inscribed in the triangle as shown in figure 2-5. In other words, the (perpendicular) distance from this point of intersection—known as the center of the inscribed circle or *incenter*—to the three sides of the triangle is the same for all three sides: $IP_a = IP_b = IP_c$, where $P_a$, $P_b$, and $P_c$ are the feet of the perpendiculars from the incenter and the points $T_a$, $T_b$, and $T_c$ are the points of intersection of the angle bisectors and the opposite side. Another way of looking at this configuration is to note that the circle is tangent to each of the three sides of the triangle. Later, we will justify that the incenter, $I$, of this circle is in fact the point of intersection of the angle bisectors. We should also note that $AT_b = AT_c$, $BT_c = BT_a$, $CT_b = CT_a$.

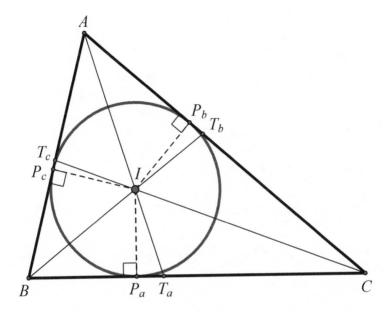

FIGURE 2-5

Every triangle has an *inscribed circle* (inside the triangle) and three *escribed circles* (outside the circle). Each of these circles is tangent to the three sides of the triangle. These are also shown in figure 2-6 for a randomly drawn triangle. These four circles are sometimes called *equicircles*. The centers $I_a$, $I_b$, and $I_c$ of the escribed circles are determined by the intersection of two external-angle bisectors and one internal-angle bisector.

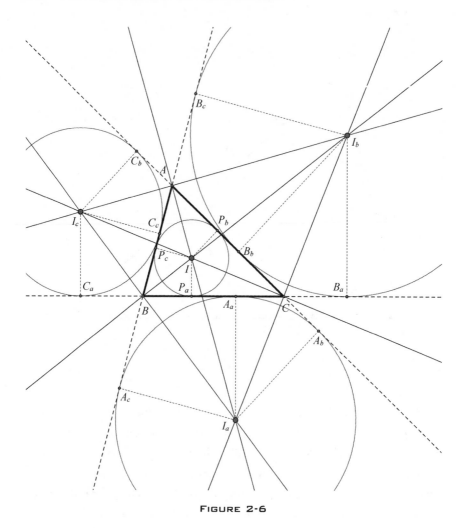

FIGURE 2-6

Once again, we have a neat relationship determined by the feet of the perpendiculars ($P_a$, $P_b$, and $P_c$) from the incenter to the sides of the triangle. Using the designations shown in figure 2-7, namely that

$$a_1 = BP_a \, , a_2 = P_aC \, , b_1 = CP_b \, , b_2 = P_bA \, , c_1 = AP_c \, , c_2 = P_cB,$$

we get $a_1^2 + b_1^2 + c_1^2 = a_2^2 + b_2^2 + c_2^2$ .

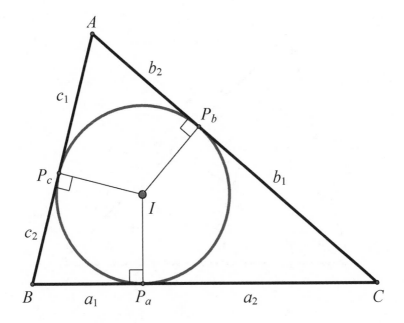

FIGURE 2-7

## INTRODUCTION TO THE MEDIANS OF A TRIANGLE

The medians of a triangle, which join a vertex of a triangle with the midpoint of the opposite side, trisect each other at their point of concurrency. That is, in figure 2-8 the following is true:

$$AG = 2 \cdot GM_a$$
$$BG = 2 \cdot GM_b$$
$$CG = 2 \cdot GM_c$$

Furthermore, the point $G$ is the center of gravity, or the *centroid*, of the triangle—that is, the balancing point. If you have a cardboard triangle, and you want to balance it on the point of a pencil, then the point at which it would balance is the centroid.

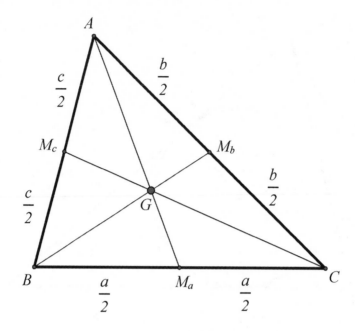

**FIGURE 2-8**

We can prove the mutual trisections of the three medians through the geometry taught in high school. We will show one possible method here. However, the power of knowing some geometric theorems beyond those taught in high school will come in very handy, as we shall show a bit further on. We will show a very simple procedure using elementary methods.

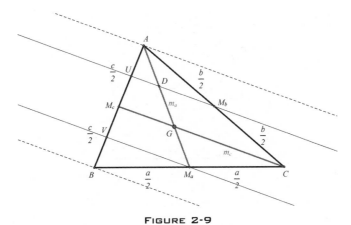

FIGURE 2-9

We begin by drawing parallel lines to $CM$, and going through points $A, M_b, M_a, B$, as shown in figure 2-9.

For triangle $ACM_c$ we have $AM_b = CM_b$; therefore, $AU = UM_c$. In a similar fashion, we can eventually show that $AU = UM_c = \dfrac{c}{4}$, and $BV = VM_c = \dfrac{c}{4}$.

We have $AU = UM_c = M_cV$, and it follows that $AD = DG = GM_a = \dfrac{m_a}{3}$. This relationship can be simply repeated for each of the other two medians, which in turn would allow us to establish concurrency at the trisection point—the centroid.

## INTRODUCTION TO THE PERPENDICULAR BISECTORS OF A TRIANGLE

The perpendicular bisector $p_c$ of the line segment $AB$ is a line that is perpendicular to $AB$ and contains its midpoint $M_c$ (see figure 2-10). Every point, $P$, along the perpendicular bisector of $AB$ is equidistant from the endpoints of $AB$, so that $AP = BP$. The three perpendicular bisectors of a triangle $ABC$, $p_a, p_b$, and $p_c$, are concurrent at the point $O$, which is the center of the circumscribed circle of triangle $ABC$.

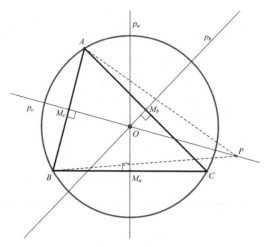

FIGURE 2-10

This can be easily justified in that the point $O$ is the intersection of $p_a$ and $p_b$ (see figure 2-11). They form the property of equidistance from the endpoints, so we get (for $p_a$) $BO = CO$ and (for $p_b$) $AO = CO$. Therefore, $AO = BO$. Consequently, point $O$ must also be on the perpendicular bisector $p_c$. Thus, the three perpendicular bisectors are concurrent at point $O$.

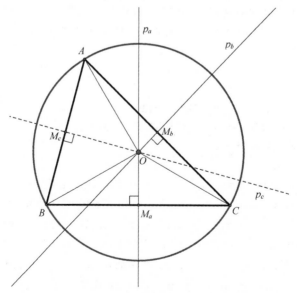

FIGURE 2-11

# INTRODUCTION TO CEVA'S THEOREM:
# A SIMPLER WAY OF
# PROVING CONCURRENCY

When the altitudes, angle bisectors, and medians are presented in a high-school geometry course, the proofs of their concurrency that are provided are not at all simple. These proofs are often bypassed for convenience. Yet, if one delves into some geometry beyond what is presented in high school, there is a theorem that makes proving these three concurrencies much simpler. This very powerful and useful theorem was first published by Giovanni Ceva (1647–1734) in 1678 in his *De lineis invicem secantibus statica construction*. Ceva's theorem is an equivalence—or biconditional—which means that the converse is also true. To justify it requires two proofs—the original statement and its converse. For us to accept this theorem it would be appropriate to prove its validity. Ceva's theorem states the following:

The three lines containing the vertices of triangle *ABC* (figure 2-12) and intersecting the opposite sides in points *L*, *M*, and *N*, respectively, are concurrent if and only if $\frac{AM}{MC} \cdot \frac{BN}{NA} \cdot \frac{CL}{BL} = 1$.

There are many proofs available to justify this theorem, yet we shall use just one of these methods to prove this wonderful theorem. It is perhaps easier to follow the proof looking at the left-side diagram and then verifying the validity of each of the statements in the right-side diagram. In any case, the statements made in the proof hold for *both* diagrams.

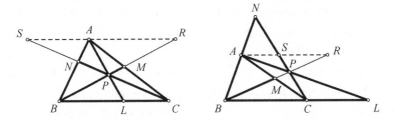

FIGURE 2-12

To prove the "only if" part of the theorem, we consider figure 2-12, where we have on the left triangle $ABC$ with a line $(SR)$ containing $A$ and parallel to $BC$, intersecting $CP$ extended at $S$ and $BP$ extended at $R$.

The parallel lines enable us to establish the following pairs of similar triangles:

$$\triangle AMR \sim \triangle CMB, \text{ therefore } \frac{AM}{MC} = \frac{AR}{CB} \tag{I}$$

$$\triangle BNC \sim \triangle ANS, \text{ therefore } \frac{BN}{NA} = \frac{CB}{SA} \tag{II}$$

$$\triangle CLP \sim \triangle SAP, \text{ therefore } \frac{CL}{SA} = \frac{LP}{AP} \tag{III}$$

$$\triangle BLP \sim \triangle RAP, \text{ therefore } \frac{BL}{RA} = \frac{LP}{AP} \tag{IV}$$

From (III) and (IV) we get: $\dfrac{CL}{SA} = \dfrac{BL}{RA}$

This can be rewritten as $\dfrac{CL}{BL} = \dfrac{SA}{RA}$ $\tag{V}$

Now by multiplying (I), (II), and (V) we obtain our desired result:

$$\frac{AM}{MC} \cdot \frac{BN}{NA} \cdot \frac{CL}{BL} = \frac{AR}{CB} \cdot \frac{CB}{SA} \cdot \frac{SA}{RA} = 1 .$$

This can also be written as $AM \cdot BN \cdot CL = AN \cdot BL \cdot CM$. A nice way to read this theorem is that the product of the alternate segments along the sides of the triangle, made by the concurrent line segments, or *cevians*, emanating from the triangle's vertices and ending at the opposite side, are equal. (A line segment joining the vertex of a triangle with a point on the opposite side is called a *cevian*.)

Yet it is the converse of this statement that is of particular value to use here. That is, if the products of the alternate segments along the sides of the triangle are equal, then the cevians determining these points must be concurrent.

We shall now prove that, if the lines containing the vertices of triangle $ABC$ intersect the opposite sides in points $L, M,$ and $N$, respectively, so that $\frac{AM}{MC} \cdot \frac{BN}{NA} \cdot \frac{CL}{BL} = 1$, then these lines $AL, BM,$ and $CN$ are concurrent.

Suppose *BM* and *AL* intersect at *P*. Draw *PC* and call its intersection with *AB*, point *N'*. Now that *AL*, *BM*, and *CN'* are concurrent, we can use the part of Ceva's theorem proved earlier to state the following:

$$\frac{AM}{MC} \cdot \frac{BN'}{N'A} \cdot \frac{CL}{BL} = 1.$$

But our hypothesis stated that $\frac{AM}{MC} \cdot \frac{BN}{NA} \cdot \frac{CL}{BL} = 1$.

Therefore, $\frac{BN'}{N'A} = \frac{BN}{NA}$, so that *N* and *N'* must coincide, and thereby proves the concurrency.

For convenience, we can restate this relationship as:

If $AM \cdot BN \cdot CL = MC \cdot NA \cdot BL$, then the three lines are concurrent.

There is an interesting variation to the famous Ceva theorem, one discovered by the French mathematician Lazare Carnot (1753–1823). Here the concurrency of the three cevians is specified by the partitioned angles at the triangles vertices. In figure 2-13, we have cevians *AL*, *MB*, and *CN* that partition the angles as angle *A* into $\alpha_1$ and $\alpha_2$, angle *B* into $\beta_1$ and $\beta_2$, and angle *C* into $\gamma_1$ and $\gamma_2$.

They will meet at a common point *P* if and only if[5] $\frac{\sin \alpha_1}{\sin \alpha_2} \cdot \frac{\sin \beta_1}{\sin \beta_2} \cdot \frac{\sin \gamma_1}{\sin \gamma_2} = 1$. The proof entails using the law of sines several times and can be found in various geometry books.[6]

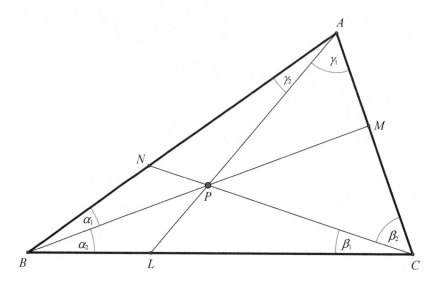

FIGURE 2-13

## USING CEVA'S THEOREM TO PROVE SOME FAMILIAR CONCURRENCIES

As we mentioned earlier, Ceva's theorem is not typically a part of the high-school course, yet it provides us with a very powerful—yet simple—tool for proving concurrencies. Follow along as we now prove the basic concurrencies that we encountered earlier, yet with very short proofs—using Ceva's theorem.

### CONCURRENCY OF THE ALTITUDES OF A TRIANGLE

We can use Ceva's theorem to prove that the altitudes of triangle $ABC$ (see figure 2-2) are concurrent.

In triangle $ABC$, where $AH_a$, $BH_b$, and $CH_c$ are altitudes:[7]

$$\triangle AH_c C \sim \triangle AH_b B, \text{ so that } \frac{AH_c}{H_b A} = \frac{AC}{AB}. \tag{I}$$

$\triangle BH_aA \sim \triangle BH_cC$, so that $\dfrac{BH_a}{H_cB} = \dfrac{AB}{BC}$. \hfill (II)

$\triangle CH_bB \sim \triangle CH_aA$, so that $\dfrac{CH_b}{H_aC} = \dfrac{BC}{AC}$. \hfill (III)

Multiplying (I), (II), and (III) gives us

$$\frac{AH_c}{H_bA} \cdot \frac{BH_a}{H_cB} \cdot \frac{CH_b}{H_aC} = \frac{AC}{AB} \cdot \frac{AB}{BC} \cdot \frac{BC}{AC} = 1,$$

or $AH_c \cdot BH_a \cdot CH_b = H_bA \cdot H_cB \cdot H_aC.$

By Ceva's theorem, this indicates that the altitudes are concurrent.

## CONCURRENCY OF THE
## ANGLE BISECTORS OF A TRIANGLE

To prove that the angle bisectors are concurrent using Ceva's theorem, we must rely on a relationship about angle bisectors introduced in the high-school geometry course, but also proved in the appendix, namely, that the angle bisector of a triangle divides the side to which it is drawn proportional to the two adjacent sides.

In figure 2-5, for angle bisector $AT_a$, we get $\dfrac{AC}{AB} = \dfrac{CT_a}{T_aB}$.

For angle bisector $BT_b$, we get $\dfrac{AB}{BC} = \dfrac{AT_b}{T_bC}$.

For angle bisector $CT_c$, we get $\dfrac{BC}{AC} = \dfrac{BT_c}{T_cA}$.

Now multiplying these three equations, we get

$$\frac{CT_a}{T_aB} \cdot \frac{AT_b}{T_bC} \cdot \frac{BT_c}{T_cA} = \frac{AC}{AB} \cdot \frac{AB}{BC} \cdot \frac{BC}{AC} = 1,$$

which (by Ceva's theorem) allows us to conclude that the angle bisectors are concurrent.

## CONCURRENCY OF THE MEDIANS OF A TRIANGLE

The proof that the medians are concurrent using Ceva's theorem is the simplest of all. We can see from the fact that each side is divided into two equal segments that the products of the alternate segments will be equal. Therefore, the medians must be concurrent. That can be seen in figure 2-14 (left side), where $AM_c = M_cB$, $BM_a = M_aC$, and $CM_b = M_bA$.

Since $AM_c \cdot BM_a \cdot CM_b = M_cB \cdot M_aC \cdot M_bA$, we can conclude (by Ceva's theorem) that the three medians are concurrent.

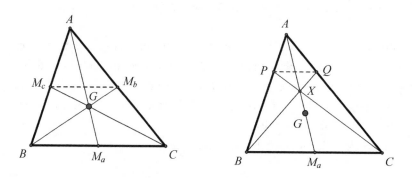

FIGURE 2-14

## EXPLORING OTHER CONCURRENCIES

We are just beginning to scratch the surface of the almost-boundless concurrencies that can be found in a triangle. To begin our exploration, let's consider the medians of triangle *ABC*, as shown in figure 2-14 (left side), and notice that $M_cM_b$ is parallel to $BC$. Suppose we slide $M_cM_b$ toward point $A$, but keeping it parallel to $BC$, where $P$ and $Q$ are now the points of intersection with $AB$ and $AC$, respectively, as shown in figure 2-14 (right side). We can show—using Ceva's theorem—that the three lines, $BQ$, $CP$, and $AM_a$ are also concurrent (at $X$).

We already have $PQ\|BC$.

Therefore, $\frac{AP}{PB} = \frac{AQ}{QC}$ or $\frac{AP}{PB} \cdot \frac{QC}{AQ} = 1.$    (I)

Since $AM_a$ is a median, $BM_a = M_aC$. Therefore, $\frac{BM_a}{M_aC} = 1.$    (II)

By multiplying the equations marked (I) and (II), we get

$$\frac{AP}{PB} \cdot \frac{QC}{AQ} \cdot \frac{BM_a}{M_aC} = 1.$$

Thus by Ceva's theorem, $AM_a$, $QB$, and $PC$ are concurrent.

As we continue to explore concurrencies in a triangle, we shall consider the following very simple, yet surprising, relationship. The lines joining each of the vertices of a triangle and the point of tangency of the inscribed circle with the opposite side are concurrent. (See figure 2-15.) This very simple relationship was first published by the French mathematician Joseph-Diaz Gergonne (1771–1859). Gergonne reserved a distinct place in the history of mathematics as the initiator in 1810 of the first purely mathematical journal, *Annales des mathématiques pures et appliqués*. The journal appeared monthly until 1832 and was known as *Annales del Gergonne*. During the time of its publication, Gergonne had published about two hundred papers, mostly on geometry. Gergonne's *Annales* played an important role in the establishment of projective and algebraic geometry, as it gave some of the greatest minds of the times an opportunity to share information. We shall consider this rather simple theorem established by Gergonne, as it exhibits concurrency and is easily proved using Ceva's theorem in the following way.

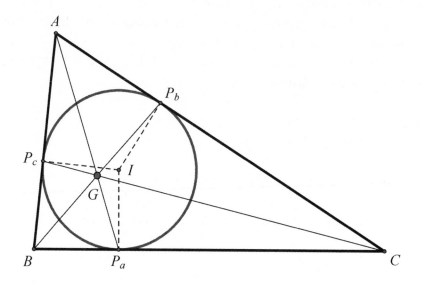

FIGURE 2-15

In figure 2-15, circle $I$ is tangent to sides $BC$, $AC$, and $AB$ at points $P_a$, $P_b$, and $P_c$, respectively. It follows that $AP_b = AP_c$, $BP_a = BP_c$, and $CP_a = CP_b$. Each of these equalities may be written as

$$\frac{AP_c}{AP_b} = 1, \frac{BP_a}{BP_c} = 1, \text{ and } \frac{CP_b}{CP_a} = 1.$$

By multiplying these three fractions we get

$$\frac{AP_c}{AP_b} \cdot \frac{BP_a}{BP_c} \cdot \frac{CP_b}{CP_a} = 1 \text{ or } AP_c \cdot BP_a \cdot CP_b = AP_b \cdot BP_c \cdot CP_a.$$

Having now established that $\frac{AP_c}{BP_c} \cdot \frac{BP_a}{CP_a} \cdot \frac{CP_b}{AP_b} = 1$, by applying Ceva's theorem, we can conclude that $AP_a$, $BP_b$, and $CP_c$ are concurrent at $G$. This point of concurrency, $G$, is called the *Gergonne point* of triangle $ABC$.

Suppose we now draw the triangle formed by joining the points of tangency of the inscribed circle. This triangle, $\Delta P_a P_b P_c$, is often called a *Gergonne triangle*. Naturally, as for all triangles, the Gergonne triangle has many "centers." One of these centers, or points of concurrency, is special in that it ties the Gergonne triangle to the original triangle. If we locate the three midpoints, $M_a$, $M_b$, and $M_c$, of the sides of the original triangle $ABC$, and from these points we draw perpendicular lines, $DM_a$, $EM_b$, and $FM_c$, to the sides of the Gergonne triangle, surprisingly, we find that these three perpendiculars are concurrent at the point $P$, as may be seen in figure 2-16.

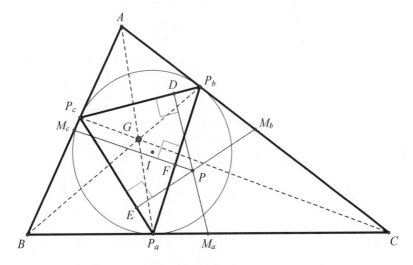

FIGURE 2-16

We can now even take this a step further and find some more concurrency points in this geometric configuration. As in figure 2-17, instead of the midpoints ($M_a$, $M_b$, $M_c$) of the sides of the original triangle, we locate the midpoints $D$, $E$, and $F$ of the circumscribed circle's arcs $\overset{\frown}{BC}$, $\overset{\frown}{CA}$, and $\overset{\frown}{AB}$, respectively. Then we draw the lines joining these arc-midpoints with the respective tangency points $P_a$, $P_b$, and $P_c$ of the inscribed circle with the triangle—which are the vertices of the Gergonne triangle. Surprisingly, we find these three lines ($DP_a$, $EP_b$, and $FP_c$) are concurrent at point $Q$. As an extra feature, we also notice that this point of concurrency ($Q$) is collinear (lies on the same line) with the center ($I$) of the inscribed circle and the center ($O$) of the circumscribed circle of the original triangle. The surprises keep on coming!

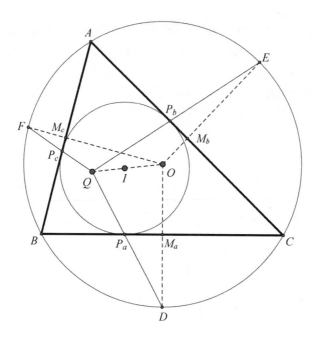

**FIGURE 2-17**

As if this were not enough, we also notice that there is another concurrency point, which we will find in the following manner. We begin by drawing the diameters of the inscribed circle from each of the three points, $P_a$, $P_b$, and $P_c$, of tangency (i.e., perpendicular to the sides) to meet the inscribed circle (whose center is the intersection of the perpendiculars at points $P_a$, $P_b$, and $P_c$) at points $D$, $E$, and $F$. Then, by drawing the lines joining each of these three points with the nearest vertex point of the original triangle (as shown in figure 2-18), we once again get a concurrency of lines. This time lines $AE$, $BD$, and $CF$ meet at point $S$.

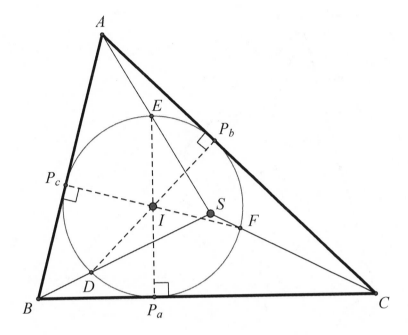

FIGURE 2-18

When we drew these diameters, there was some built-in symmetry and so the result was somewhat plausible. However, if we take this another step further and draw any three concurrent lines emanating from the tangency points ($P_a$, $P_b$, and $P_c$) with the inscribed circle, we amazingly get a similar concurrency just as before. That is, in figure 2-19, when $DP_b$, $EP_a$, and $FP_c$ are concurrent at $R$, we have $AE$, $BD$, and $CF$ concurrent at point $T$ (see figure 2-19).

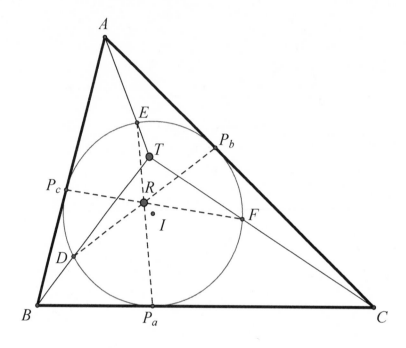

FIGURE 2-19

To further show how concurrency seems to emerge in unexpected places, we consider triangle $ABC$ with the three concurrent (at point $P$) cevians $AD$, $BE$, and $CF$, from which we then draw triangle $DEF$ and its inscribed circle. The points of tangency are, $D'$, $E'$, and $F'$, as shown in figure 2-20. Curiously enough, when we join these latter points ($D'$, $E'$, and $F'$) to each of their nearest vertices of triangle $ABC$, we once again have three lines that are concurrent—namely, $AD'$, $BE'$, and $CF'$ are concurrent at point $P'$.

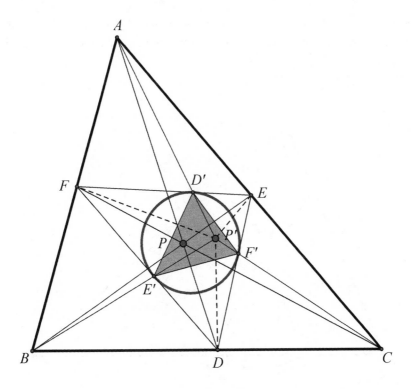

FIGURE 2-20

## SOME CONCYCLIC POINTS

It is well known that three noncollinear points determine a unique circle. When more than three points lie on the same circle we have a set of points that we tend to cherish, and we call them *concyclic points*. That is what happens when we draw lines through the Gergonne point, $G$, of a triangle and parallel to the sides of the Gergonne triangle as shown in figure 2-21, where $DE \parallel P_c P_b$, $FJ \parallel P_c P_a$, and $HK \parallel P_b P_a$.

Quite unexpectedly, where these parallel lines intersect the sides of the triangle $ABC$ is where we have the six points that lie on the same circle. Moreover, much more unexpectedly, that circle is concentric (shares the same center) with the inscribed circle of the triangle. This remarkable occurrence was first discovered by the German mathematician Carl Adams (1811–1849) in 1843[8] and, therefore, in his honor carries his name as the *Adams' circle.*[9]

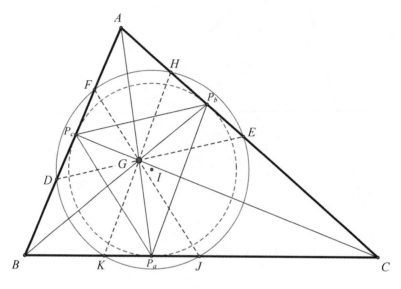

**FIGURE 2-21**

## MORE CONCURRENCIES

Recall that we arrived at the Gergonne triangle by using the tangency points of the inscribed circle of a triangle. In figure 2-22, we have the three escribed circles of triangle *ABC*. Each of these circles is tangent to the three sides of the triangle, yet each of these circles lies outside the triangle. Once again, as with the Gergonne point, the lines joining the tangency points with the opposite vertex of the triangle are concurrent at point *N*. So you can see there is a natural analog between the inscribed circle of a triangle and the escribed circles of the same triangle—in this case tied together by the Gergonne-point property.

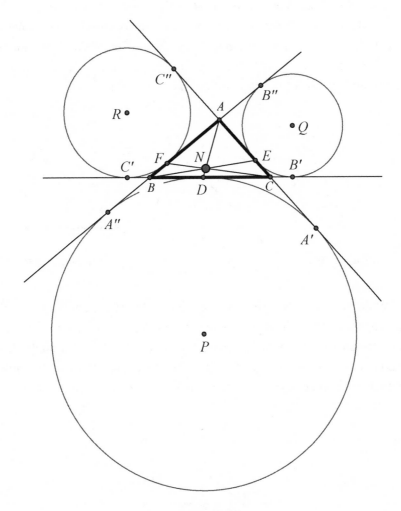

FIGURE 2-22

This is not different from the point discovered by the German mathematician Christian Heinrich von Nagel (1803–1882). We can locate this *Nagel point*, *N*, of triangle *ABC*, in the following way:

Point *P* is situated on *BC* so that *AB* + *BP* = *AC* + *CP* (see figure 2-23).

Point *Q* is situated on *AC* so that *BC* + *CQ* = *AB* + *AQ* (see figure 2-24).

Point $R$ is situated on $AB$ so that $BC + BR = AC + AR$ (see figure 2-25).

We then conclude that $AP$, $BQ$, and $CR$ are concurrent. This point of concurrency is known as the *Nagel point*, $N$, of triangle $ABC$. We should note that this is analogous to the way we located the point determined by the tangency points of the escribed circles in figure 2-22.

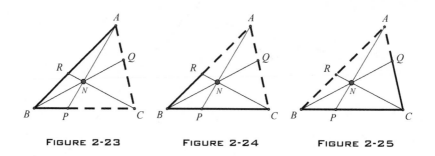

FIGURE 2-23          FIGURE 2-24          FIGURE 2-25

While we are considering the escribed circles of a triangle, we should also appreciate that they give us another point, $M$, of concurrency, one that is often called the *middle point* (or *mittenpunkt*) of a triangle. It is determined by the three lines that join the centers of the escribed circles with the midpoint of the side of the triangle to which they are tangent.

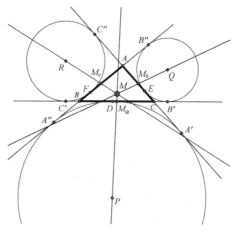

FIGURE 2-26

In figure 2-26, we have points $M_a$, $M_b$, and $M_c$ as the midpoints of the sides of triangle $ABC$. The lines $PM_a$, $QM_b$, and $RM_c$ are concurrent at point $M$, which is the middle point of the triangle $ABC$. This middle point, $M$, is not identical with the Nagel point, $N$, of the triangle $ABC$.

## CONCURRENCIES DETERMINING OTHER CONCURRENCIES

We now embark on another path toward exposing more secrets embedded in triangles. We will begin by drawing any three cevians that are concurrent and we will use this point of concurrency to find another concurrency point. This not only is unexpected but also shows how intertwined some characteristics of triangles can be.

We begin with triangle $ABC$, where we draw the cevian lines $AD$, $BF$, and $CE$ so that they are concurrent at point $P$, as shown in figure 2-27. We then consider the circle drawn through the points $D$, $E$, and $F$, which also intersects the triangle in points $D'$, $E'$, and $F'$. Using Ceva's theorem, it can be shown that the lines $AD'$, $BF'$, and $CE'$ are also concurrent (at $Q$).

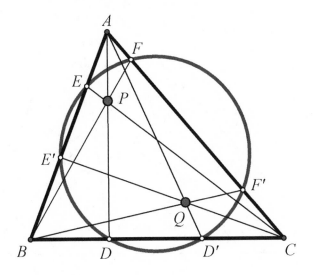

FIGURE 2-27

A somewhat analogous situation can be created as shown in figure 2-28, where we have triangle *ABC* with concurrent cevians *AL*, *BM*, and *CN* meeting at point *P*. Rather than use a circle through the feet of these cevians to determine the key points that we used above to determine another point of concurrency, we will construct parallels. We will draw *NR*∥*AC*, *LS*∥*AB*, and *MT*∥*BC* to determine points *R*, *S*, and *T*. These newly determined points will determine yet another set of concurrent cevians, namely, *AR*, *BS*, and *CT*, which meet at point *Q*.

A similar case can be made for selecting parallel lines from the feet of the original concurrent cevians (*AL*, *BM*, and *CN* ) to the other sides, namely, from point *L* parallel to *AC* instead of to *AB*, as we did earlier, from point *M* parallel to *AB* instead of *BC*, and from point *N* parallel to *BC* instead of *AC*. Actually, following this scheme, we can get lots more such points of concurrency of cevians by using each new set of concurrent cevians to generate another set of concurrent cevians. The reader may wish to explore these concurrencies further.

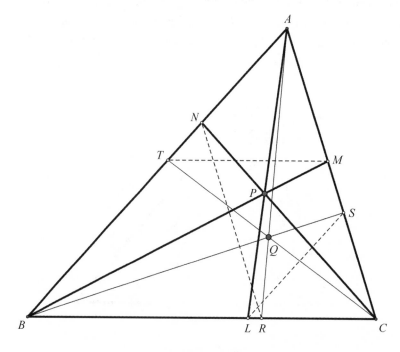

FIGURE 2-28

A similar type of reasoning will produce yet another rather unexpected concurrency. Suppose we have a circle intersecting each of the three sides of a randomly selected triangle twice, however, in such a way that the perpendiculars to the sides of the triangle at three of the points (D, E, and F) of intersection with the circle are concurrent (at P). (See figure 2-29.) When this is the case, we find that the perpendiculars to the three sides of the triangle at the remaining circle intersection points (K, L, and M) are also concurrent (at Q).

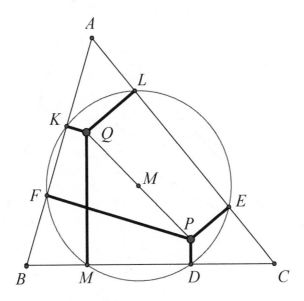

FIGURE 2-29

As if that weren't surprising enough, notice what happens when we consider the midpoint of the line segment joining these two points of concurrency. This is, in fact, the center, M, of the circle containing the six feet of perpendiculars.

We will now consider a somewhat more involved configuration—one that will consider midpoints of the sides of a randomly selected triangle, and the midpoints of a set of concurrent cevians of the triangle. These will curiously generate another set of concurrent cevians.

In triangle *ABC* (figure 2-30), the cevian lines *AL*, *BM*, and *CN* are concurrent at point *P*. Bear in mind that these cevians are *any* concurrent cevians of triangle *ABC*. We now locate the midpoints of *AL*, *BM*, and *CN* and label them as points *D*, *E*, and *F*, respectively. We also have points $M_a$, $M_b$, and $M_c$ as the midpoints of the sides of triangle *ABC*. Using Ceva's theorem, it can be shown that, quite unexpectedly, $DM_a$, $EM_b$, and $FM_c$ are also concurrent—at the point *Q*.

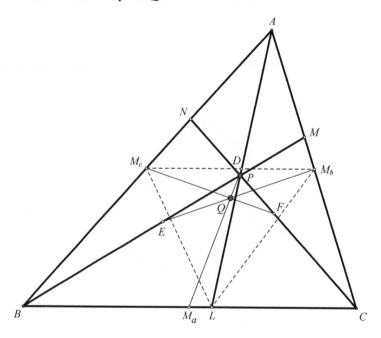

FIGURE 2-30

Having one set of concurrent cevians generating another set of concurrent cevians is always a striking curiosity. Consider the triangle *ABC* where the cevians *AL*, *BM*, and *CN* are concurrent at point *P* as shown in figure 2-31. The points *D*, *E*, and *F* are the midpoints of the line segments *LM*, *MN*, and *NL*, respectively. Surprisingly (and with the help of Ceva's theorem), the lines *AD*, *BE*, and *CF* can be shown to be concurrent at point *Q*. The triangle *DEF* will help the ambitious reader to justify this remarkable relationship.

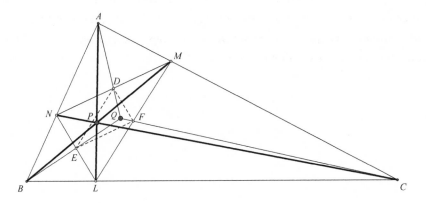

FIGURE 2-31

To generate another unexpected concurrency we can extend the previous scheme. Again, we shall begin with triangle $ABC$, where the cevians $AL$, $BM$, and $CN$ and concurrent at point $P$. Now, as shown in figure 2-32, suppose that the points $D$, $E$, and $F$ are not (necessarily) the midpoints of $NM$, $LN$, and $LM$, respectively, but rather just three points on the triangle that determine concurrent cevian lines, $LD$, $ME$, and $NF$ at point $R$. Astonishingly, these three points, $D$, $E$, and $F$, also yield similar concurrent lines as we had in the previous configuration. Namely, the lines $AD$, $BE$, and $CF$ can be shown to also be concurrent at point $Q$, using Ceva's theorem.

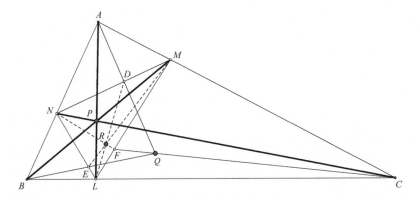

FIGURE 2-32

Concurrency in a triangle seems to be endless. There are many more such concurrencies to be discovered, and we will encounter many more in the following chapters. However, the ambitious reader may want to seek other concurrencies at this point before reading on.

# CHAPTER 3

# NOTEWORTHY POINTS
# IN A TRIANGLE

U p until now, the concurrencies we have exhibited have determined some significant points in a triangle.

The concurrency point of the angle bisectors of a triangle gives us the center of the inscribed circle of the triangle. The point of intersection of the perpendicular bisectors of the sides of a triangle gives us the center of the circumscribed circle of a triangle. The point of concurrency of the altitudes of a triangle gives us the orthocenter of a triangle. The point of intersection of the medians of a triangle gives us the centroid, or center of gravity, of a triangle. As we seek further significant points of a triangle, we can list the point in a triangle where the sides subtend (i.e., determined by being opposite) equal angles. For example, in figure 3-1 the point $P$ is so situated that $\angle APB = \angle BPC = \angle CPA$. This, unexpectedly, is also the very same point in a triangle from which the sum of the distances to the vertices is the smallest. That is, where $AP + BP + CP$ is smaller than the sum of the distances from any other point in the triangle to the vertices. These are just two properties of a very special point in a triangle that will provide us with some other quite-surprising results. Follow along!

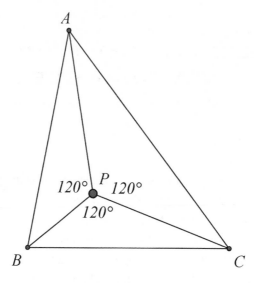

FIGURE 3-1

As we now begin this rather interesting exploration of this particularly significant point in a triangle, we consider triangle *ABC* on whose sides we shall construct equilateral triangles as shown in figure 3-2. Then we will draw the lines *AA'*, *BB'*, and *CC'*. Using some elementary geometry, we can easily show that these three line segments are equal. This is done by proving pairs of triangles congruent. Now we say "easily," yet most high-school students who attempt to prove this equality — usually given as an exercise with congruence proofs early in the course — have difficulty doing it, simply because they have difficulty identifying the triangles that have to be proved congruent to establish the line-segment equality. Once the triangles have been identified, the proof is quite simple. That is, to show *AA'* = *CC'*, we merely prove that $\triangle AA'B \cong \triangle BC'C$, as shown in figure 3-3 (by the SAS congruence relationship). Similarly, *AA'* = *BB'*, by showing, in similar fashion, that $\triangle AA'C \cong \triangle B'BC$.

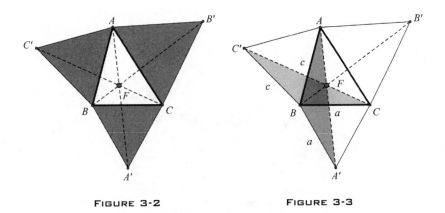

FIGURE 3-2                              FIGURE 3-3

Now having justified that the line segments $AA'$, $BB'$, and $CC'$ are equal, it is interesting to note that the lines are also concurrent, at a point called the *Fermat point*, $F$, named after the French mathematician Pierre de Fermat (1607–1665).

To prove this concurrency we would draw the circumscribed circles of the three equilateral triangles and show that they contain a common point $F$, as shown in figure 3-4.

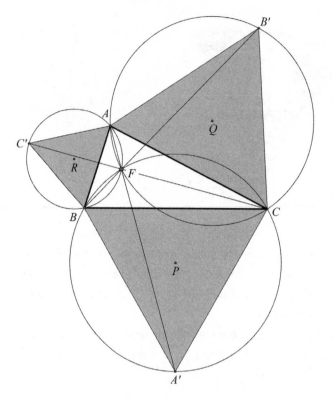

FIGURE 3-4

Let's now begin by considering the circumscribed circles of the three equilateral triangles $BCA'$, $ACB'$, and $ABC'$, where $P$, $Q$, and $R$ are the centers of these circles. We show this in figure 3-4. The points of intersection of the circles $Q$ and $R$ are points $F$ and $A$. Now we want to show that the point $F$ is also on circle $P$.

Since $\overparen{AC'B} = 240°$, we know that the inscribed angle $\angle AFB = \frac{1}{2}\left(\overparen{AC'B}\right) = 120°$.

Similarly, $\angle AFC = \frac{1}{2}\left(\overparen{AB'C}\right) - 120°$. Therefore $\angle BFC = 120°$, since a complete revolution is $360°$.

Since arc $BA'C = 240°$, $\angle BFC$ is an inscribed angle and point $F$ must, therefore, lie on circle $P$. Thus, we can see that the three circles are concurrent, intersecting at point $F$.

Now by joining point $F$ with points $A$, $B$, $C$, $A'$, $B'$, $C'$, we find that $\angle B'FA = \angle AFC' = \angle C'FB = 60°$, and therefore $B'FB$ is a straight line. Similarly $C'FC$ and $A'FA$ are also straight lines, which establishes the concurrency of the lines $AA'$, $BB'$, and $CC'$. In this way, we can determine the point in triangle $ABC$ at which the three sides subtend congruent angles. The point $F$ is also called the *equiangular point* of triangle $ABC$ since $\angle AFB = \angle AFC = \angle BFC = 120°$.

## NAPOLEON'S THEOREM

At this point we are ready to embark on a rather famous theorem in geometry, one that is attributed to Napoleon Bonaparte (1769–1821), who, aside from his fame in French history as a major figure for his military prowess, also distinguished himself as a topflight mathematics student in school and later at the Paris Military School and earned membership in the Institut de France, a prestigious scientific society. He prided himself on his talent in mathematics, and particularly in geometry. Yet the theorem that still bears his name today was first published by Dr. W. Rutherford in *The Ladies' Diary* in 1825, four years after Napoleon's death. To this day it is uncertain if Napoleon ever knew of the relationship we are about to investigate.

To show Napoleon's theorem, we consider the centers of the three circumscribed circles of the three equilateral triangles drawn on the sides of triangle $ABC$. (See figure 3-5.) These centers determine an equilateral triangle $PQR$, known as the *Napoleon triangle*. The Napoleon triangle can be shown to be equilateral by demonstrating that the sides of triangle $PQR$ are proportional to the equal line segments $AA'$, $BB'$, and $CC'$, thus making the three sides of triangle $PQR$ equal and establishing it as an equilateral triangle. The proof of this relationship can be found in the appendix.

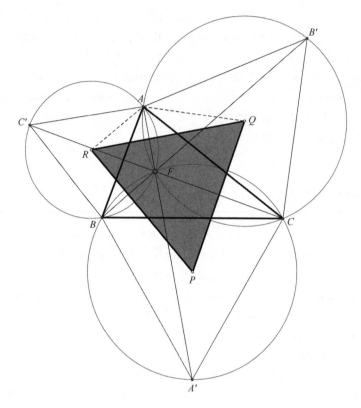

FIGURE 3-5

There are many unusual relationships tying the Napoleon triangle to the original triangle. For one, the Napoleon triangle and the original triangle share a common centroid. As we establish this interesting relationship, we will encounter some other curiosities along the way. To begin this quest, in figure 3-6 we have point $G$ as the centroid of triangle $ABC$, and point $P$ the centroid of triangle $BCA'$. We shall denote point $M_a$ as the midpoint of $BC$. Because the centroid of a triangle trisects each of the medians, we have $AM_a = 3GM_a$, and $A'M_a = 3PM_a$. Since $GP$ partitions $AM_a$ and $A'M_a$ proportionally, we can conclude that $\Delta M_a GP \sim \Delta M_a AA'$, and $AA' = 3GP$, or the distance between the centroids is one-third the length of the line segment joining a vertex of the original triangle with the remote vertex of the relevant equilateral triangle.

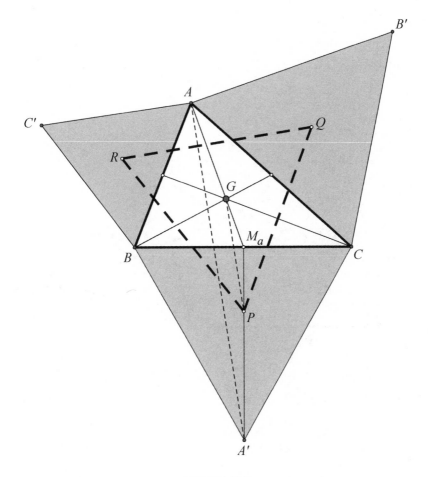

FIGURE 3-6

In a similar fashion, we can show that $CC' = 3GR$, and $BB' = 3GQ$ (see figure 3-7). We have shown earlier (pp. 66, 67) that $AA' = BB' = CC'$. Therefore, $GP = GQ = GR$. Since triangle $PQR$ is equilateral and the distances from point $G$ to its vertices are equal, we can conclude that $G$ is also the centroid of triangle $PQR$. We have shown, therefore, that point $G$ is the centroid of both the outer Napoleon triangle $PQR$ and the original triangle $ABC$.

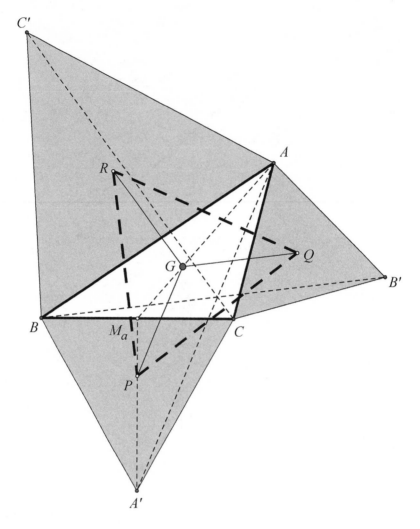

FIGURE 3-7

All that we have said so far about the three equilateral triangles drawn on the sides of a randomly selected triangle was based on their being drawn *externally* to the original triangle. Yet, as you might have expected by now, we can make all the same arguments about a configuration that has the three equilateral triangles drawn *internally* to the given triangle, or, shall we say, overlapping the original triangle, as shown in figure 3-8. We have triangle *UVW* equilateral and sharing its centroid point, *G*, with the centroids of triangle *PQR* and triangle *ABC*.

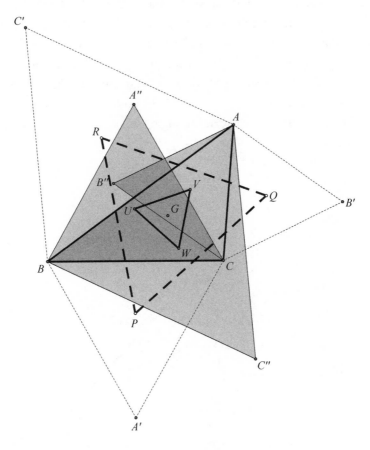

FIGURE 3-8

Referring to any of the previous few figures, say figure 3-7, suppose we now leave triangle *BCA'* fixed and move point *A* to various positions (even on the other side of *BC*). As long as the point *A* does not land on points *B* or *C*, where triangle *ABC* would then have a zero area, all that we have established above will still hold true. This is truly an amazing relationship! It can be nicely shown with dynamic geometry software, such as the Geometer's Sketchpad® or GeoGebra®.

As if the configuration shown in figure 3-2 did not already produce enough unexpected equilateral triangles, we can find yet another one. All we need to do is to construct a parallelogram $AC'CD$ as shown in figure 3-9, and we can identify an equilateral triangle, namely, $AA'D$. The same size equilateral triangle can be produced at all sides of this configuration, since each of these equilateral triangles will use for a side one of the equal lengths $AA'$, $BB'$, and $CC'$. To justify that triangle $AA'D$ is, in fact, equilateral, we can show that $AD = AA'$, since both are equal to $CC'$, and $\angle DAA' = 60°$, since it is alternate-interior to $\angle AFC' = 60°$, where $F$ is the Fermat point. That is, $AA'D$ is an isosceles triangle with a 60° vertex angle, which makes it equilateral.

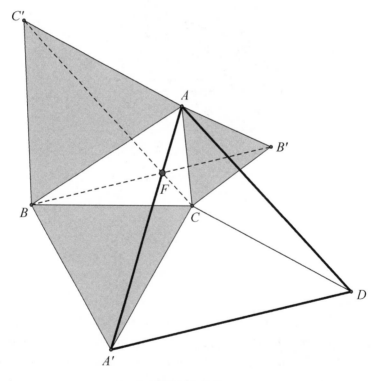

FIGURE 3-9

By connecting a vertex of each of the three equilateral triangles to the nearest vertex of the Napoleon triangle $PQR$, we find the lines concurrent at the center $O$ of the circumscribed circle of triangle $ABC$, as shown in figure 3-10.

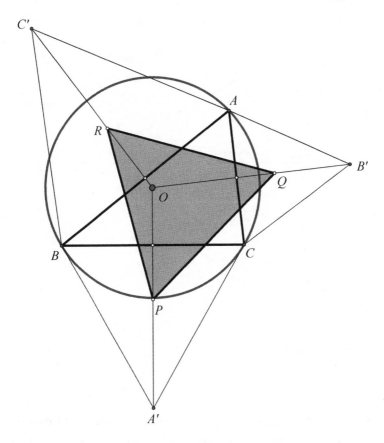

FIGURE 3-10

We are not yet finished with this rich equilateral-triangle configuration. We need to focus again on the Fermat point $F$. Not only is point $F$ the equiangular point, but, as we indicated earlier, it will also turn out to be the *minimum-distance point* from the three vertices of triangle $ABC$—that is, the sum of the distances from that point to the three vertices of the triangle is less than the sum of the distances from any other point in the triangle to the vertices. In other words, this point now has *two* important properties: the minimum-distance point and the equiangular point of the triangle.

Let's investigate how we can justify this last claim. We begin by considering triangle *ABC* with no angle measuring greater than 120°, as shown in figure 3-11.

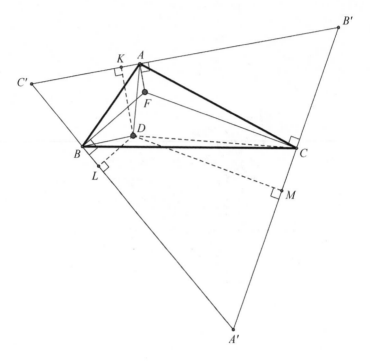

FIGURE 3-11

To show that the sum of the distances from point *F* to each of the three vertices of triangle *ABC* is less than that from any other point to the vertices, we need to take any other randomly selected point *D* and show that the sum of the distances from this point is greater than the sum of the distances from point *F* to the vertices of the triangle. The justification—or proof—of this relationship is quite interesting and a bit different from other geometric proofs. Follow along and you will find it rewarding!

Let *F* be the equiangular point in the interior of triangle *ABC*, that is, where $\angle AFB = \angle BFC = \angle AFC = 120°$.

Draw lines through *A*, *B*, and *C*, which are perpendicular to *AF*, *BF*, and *CF*, respectively. These lines meet to form yet another equilateral triangle,

$A'B'C'$. (To prove triangle $A'B'C'$ is equilateral, notice that each angle has measure 60°. This can be shown by considering, for example, quadrilateral $AFBC'$. Since $\angle C'AF = \angle C'BF = 90°$, and $\angle AFB = 120°$, it follows that $\angle AC'B = 60°$.) Let $D$ be *any other* point in the interior of triangle $ABC$. We must then show that the sum of the distances from $F$ to the vertices of triangle $ABC$ is less than the sum of the distances from the randomly selected point $D$ to the vertices of triangle $ABC$.

We can easily show that the sum of the distances from any point in the interior of an equilateral triangle to the sides is a constant, namely, the length of the altitude. Let's take a moment to review this, as we already encountered this as Viviani's theorem in chapter 1.

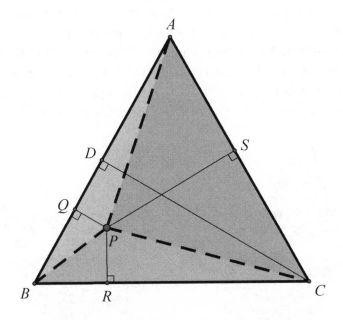

FIGURE 3-12

Consider equilateral triangle $ABC$, where $PQ \perp AB, PR \perp BC, PS \perp AC$, and $CD \perp AB$. Draw the line segments $PA, PB$, and $PC$ (see figure 3-12).

Area $\triangle ABC$ = Area $\triangle APB$ + Area $\triangle BPC$ + Area $\triangle CPA$

$=\frac{1}{2}AB \cdot PQ + \frac{1}{2}BC \cdot PR + \frac{1}{2}AC \cdot PS$.

Since $AB = BC = AC$, the Area $\triangle ABC = \frac{1}{2} AB \cdot (PQ + PR + PS)$.

However, the *Area* $\triangle ABC = \frac{1}{2} AB \cdot CD$. Therefore, the sum $PQ + PR + PS = CD$ is a constant for the given triangle.

Now using this constant relationship, we have in figure 3-11, $FA + FB + FC = DK + DL + DM$ (where $DK$, $DL$, and $DM$ are the perpendiculars to $B'KC'$, $A'LC'$, and $A'MB'$, respectively).

But $DK + DL + DM < DA + DB + DC$. (The shortest distance from an external point to a line is the length of the perpendicular segment from that point to the line.) By substitution: $FA + FB + FC < DA + DB + DC$.

You may wonder why we chose to restrict our discussion to triangles with angles of measure less than 120°. If you try to construct the point $F$ in a triangle with one angle of measure of 150°, the reason for our restriction will become obvious. In figure 3-13, we have $\angle BAC > 120°$, and we find that the supposed minimum-distance point is outside of triangle $ABC$. When $\angle BAC = 120°$, as shown in figure 3-14, the minimum-distance point is on vertex $A$. Therefore, the minimum-distance point *in* a triangle (with no angle of measure greater than 120°) is the equiangular point (i.e., the point at which the sides of the triangle subtend congruent angles).

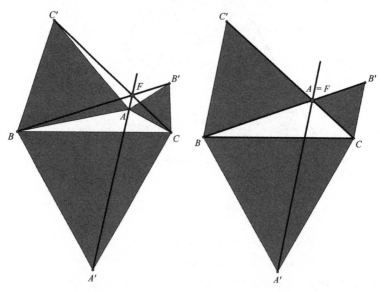

FIGURE 3-13                    FIGURE 3-14

## SQUARES ON THE SIDES OF A TRIANGLE

While there are other discoveries possible when equilateral triangles are placed on the sides of a randomly drawn triangle, as we have done in figure 3-2, there are also some startling results when one constructs a *square* on the sides of a randomly drawn triangle, as shown in figure 3-15.

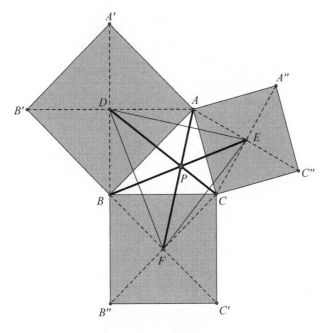

FIGURE 3-15

Our first observation involves the lines joining the centers of the squares in figure 3-15 to the remote vertices of the triangle *ABC*. As you might have expected by now, they are concurrent. Then we also find that each of these concurrent lines is perpendicular to one of the sides of the triangle formed by joining the centers of the three squares. You may also describe these three concurrent lines, *AF*, *BE*, and *CD*, as overlapping the altitudes of triangle *DEF*. Moreover, each of the concurrent lines, *AF*, *BE*, and *CD*, is the same length as the side of triangle *DEF*, to which it is perpendicular. Namely, *AF* = *DE*, *BE* = *DF*, and *DC* = *EF*. You may wish to search for more such gems in this configuration. Here are a few more to entice you as you begin your search:

$$AC^2 + AB''^2 = AB^2 + AC'^2,$$
$$A'A''^2 + BC^2 = 2 \cdot (AB^2 + AC^2),$$
$$AB''^2 + CA'^2 + BC''^2 = AC'^2 + CB'^2 + BA''^2.$$

However, be aware that triangle $DEF$ in figure 3-15 is not necessarily equilateral as you might expect. In figure 3-16, we again have squares drawn on the sides of triangle $ABC$. Points $P$ and $Q$ are selected so that $PB'BB''$ and $QC'CC''$ are parallelograms. Then, surprisingly, triangle $PAQ$ is an isosceles right triangle—that is, $AP = AQ$, and $\angle PAQ$ is a right angle.

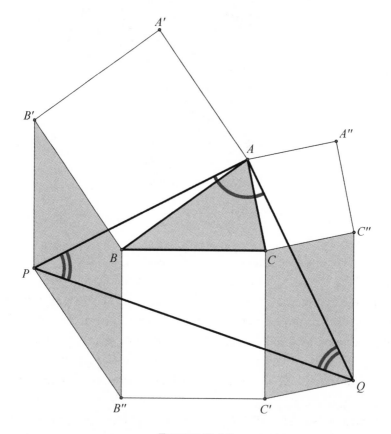

FIGURE 3-16

The Luxembourger mathematician Joseph (Jean Baptiste) Neuberg (1840–1926) discovered the following theorem, which truly depicts a noteworthy point in a triangle. In figure 3-17, we draw the three squares (externally) on the sides of the randomly drawn triangle $ABC$. The centers of these three squares, $D$, $E$, and $F$, form triangle $DEF$. We now draw the three squares on the sides of triangle $DEF$, yet overlapping triangle $DEF$. We find that the centers of these squares are the midpoints of the sides of the original triangle $ABC$. Truly a remarkable relationship of the points of a triangle!

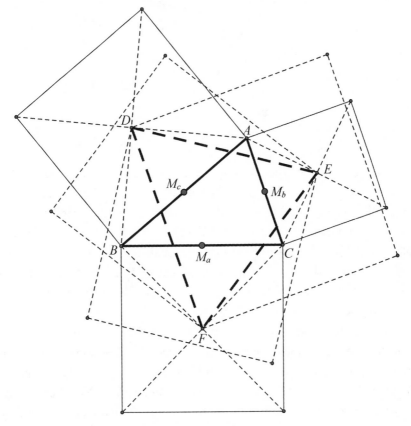

FIGURE 3-17

In order to prove this is true, we would merely need to show that the midpoint $M_a$ of $BC$ is the center of the square on $DE$, that is, that $M_aD$ and $M_aE$ are equal and perpendicular. (See figure 3-18.)

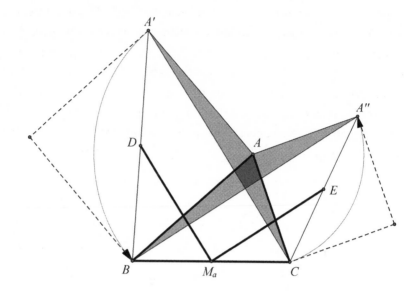

FIGURE 3-18

Rotating 90° the triangle *A'AC*, it will coincide with triangle *BAA''* (i.e., *A'* → *B*, *A* → *A*, *C* → *A''*), showing that *A'C* and *BA''* are equal and perpendicular. Since *D* and $M_a$ are the midpoints of two sides of the triangle *BA'C*, we find that $DM_a$ must be parallel to the third side, *A'C*, and half as long; similarly, $EM_a$ is parallel to *BA''* and half its length. Since *A'C* and *BA''* are equal and perpendicular, then so are $DM_a$ and $EM_a$. That means that $M_a$ is the point of intersection of the diagonals of the (overlapping) square on side *DE* of triangle *DEF*.

We can enjoy some of the truly unexpected relationships that come from this configuration—squares on the sides of a randomly drawn tri-angle—by modifying the figure a bit. In figure 3-19, we have squares drawn on the sides of a randomly drawn triangle (*ABC*). Now here comes the tricky part: we draw a triangle on the remote side of each of the squares, so that each side is parallel to a side of the original triangle (*ABC*). That is,

for Δ*A'B'L*:  *A'L*∥*BC*, *B'L*∥*AC*, *A'B'*∥*AB*;
for Δ*A''C''K*: *A''K*∥*BC*, *C''K*∥*AB*, *A''C''*∥*AC*; and
for Δ*B''C'J*:  *C'J*∥*AB*, *B''J*∥*AC*, *B''C'*∥*BC*.

When we draw the lines connecting the remote vertices of the three triangles to those of the original triangle, you will find that these three lines are also concurrent (at $Q$). This is shown in figure 3-19.

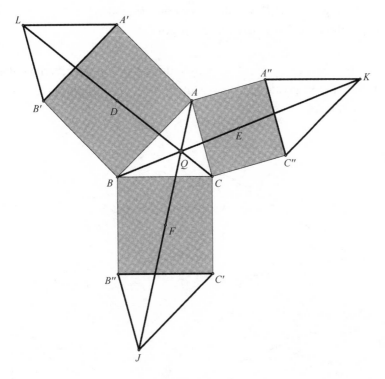

FIGURE 3-19

Now, taking the squares on the sides of a randomly drawn triangle even further, we will consider the parallelograms drawn between the squares. As shown in figure 3-20, the parallelograms are:

parallelogram $A_1AA_2A_3$ with center point $G$,
parallelogram $B_1B_3B_2B$ with center point $H$, and
parallelogram $C_1C_3C_2C$ with center point $I$.

We get concurrent lines (at $R$) here by joining the center of each parallelogram with the opposite square's center point. Again, we have a rather unexpected result.

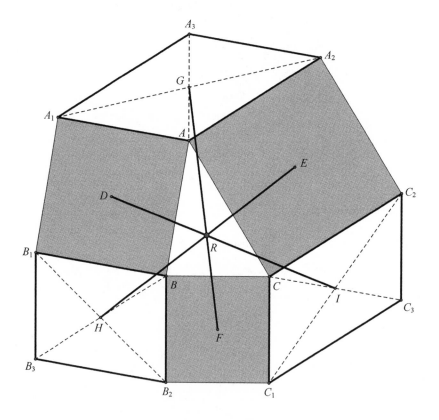

FIGURE 3-20

This configuration yields a number of additional concurrent lines, as we will see in the next several figures. In figure 3-21 we notice that the lines joining the midpoints of the sides of the original triangle with the center point of the opposite parallelogram are concurrent (at *S*).

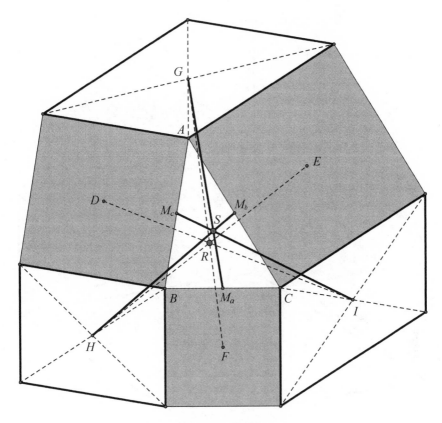

FIGURE 3-21

In figure 3-22 we notice that the diagonal extensions of the parallelograms between the squares are concurrent (at *T*). We have found, therefore, a second concurrency in the figure formed by joining the center of each square with the remote vertex of the opposite parallelogram (at *U*).

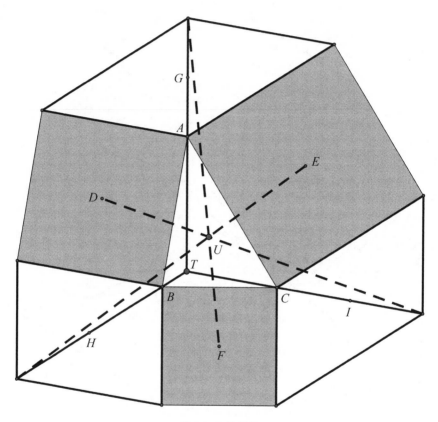

FIGURE 3-22

Yes, there are even more concurrencies on this configuration, as we will see in figure 3-23. This time we will join the midpoint of the side of each square's exterior side with the remote vertex of the opposite parallelogram. The lines are concurrent at *V*.

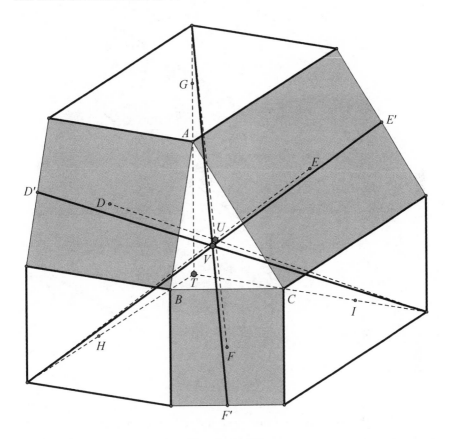

FIGURE 3-23

There is still another concurrency (shown in figure 3-24), when we join each vertex of the original triangle with the midpoint of the remote side of the square on the opposite side (at *W*).

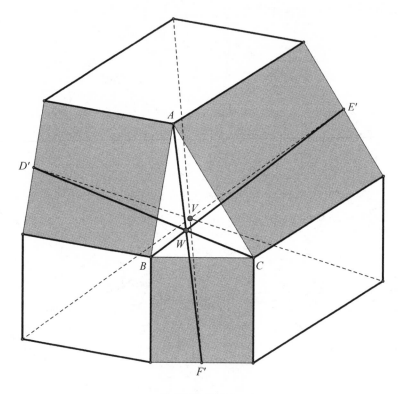

FIGURE 3-24

Although we have exhibited quite a few concurrencies in this configuration, there are many more such relationships to be found. We encourage you to search for some of them. Using a dynamic geometry program, such as the Geometer's Sketchpad or GeoGebra, can be helpful in this search. The ambitious reader will be enchanted to find that on his website "The Encyclopedia of Triangle Centers—ETC" (http://faculty. evansville.edu/ck6/encyclopedia/ETC.html) Clark Kimberling has located well over 3,600 triangle centers and shows how they can be found. These are the points of concurrency of three triangle-related lines and may serve as further motivation to search for other points of concurrency.

# CHAPTER 4

# CONCURRENT CIRCLES
# OF A TRIANGLE

A s a companion to the equiangular point (or the Fermat point, $F$) we have just discussed, there are two significant points in a triangle made popular by the French mathematician Henri Brocard (1845–1922). Before we can locate these points we must first establish how to construct a circle (of course, using only an unmarked straightedge and a compass) containing a given point and tangent to a given line (not containing that point), since we will be using this construction to locate this new point.

## A NECESSARY CONSTRUCTION

Consider the point $A$ and the line $l$, as shown in figure 4-1. We will show how to construct a circle through point $A$ and tangent to line $l$. First, we construct a perpendicular line to $l$ at the planned point of tangency, $B$. Then, we construct the perpendicular bisector of $AB$ to meet the previous perpendicular at point $O$. We can then draw the circle with center $O$ and radius $OA$, which gives us our required circle. This construction will now be used several times to locate some truly amazing points in a triangle, known as the *Brocard points*.

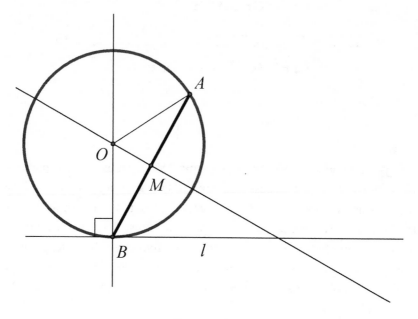

FIGURE 4-1

## BROCARD POINTS

We are now ready to locate a triangle's Brocard points. Consider the following:

In figure 4-2 you will find the three circles constructed as we did above. Each is tangent to a side of the triangle *ABC* and contains a vertex of the triangle. That is,

circle *P* is tangent to side *AC* at point *C* and contains vertex *B*,
circle *Q* is tangent to side *AB* at point *A* and contains vertex *C*, and
circle *R* is tangent to side *BC* at point *B* and contains vertex *A*.

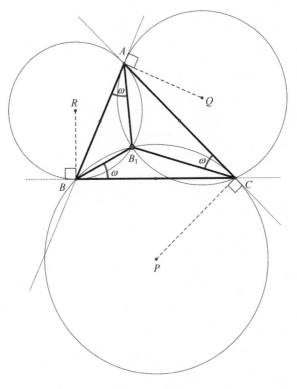

FIGURE 4-2

Two amazing things occur in this construction: First, the three circles meet at a common point, the first Brocard point, $B_1$, and, second, the angles made from this point of concurrency have the following relationship: $\angle B_1AB = \angle B_1BC = \angle B_1CA = \omega$.

A second Brocard point, $B_2$, can also be found in this triangle $ABC$ by taking the point of concurrency of these circles (see figure 4-3):

circle $S$ is tangent to side $AB$ at point $B$ and contains vertex $C$,

circle $T$ is tangent to side $BC$ at point $C$ and contains vertex $A$, and

circle $U$ is tangent to side $AC$ at point $A$ and contains vertex $B$.

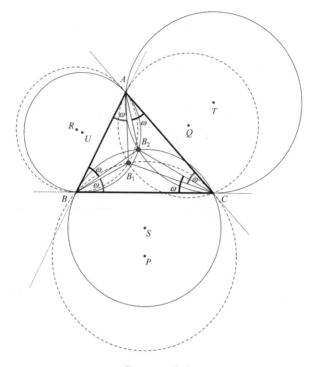

FIGURE 4-3

The Brocard points lead to lots of further advanced relationships that would be a bit beyond the scope of this book, yet they can be appreciated, nevertheless, for the unusual properties they present.

There is also a time when one concurrency generates another concurrency, as we have seen earlier. In figure 4-4, the three lines $AP$, $BP$, and $CP$ are concurrent at point $P$. If we now take three line segments that make equal angles with the adjacent sides of the angle of the triangle so that

$$\angle QAC = \angle PAB \, (= \rho), \angle QBC = \angle PBA \, (= \sigma), \text{ and } \angle QCA = \angle PCB \, (= \tau),$$

then $QA$, $QB$, and $QC$ are also concurrent. This quite-surprising result is one that would not be easily anticipated yet continues to exhibit the secrets of the beauties that are embedded in triangles.

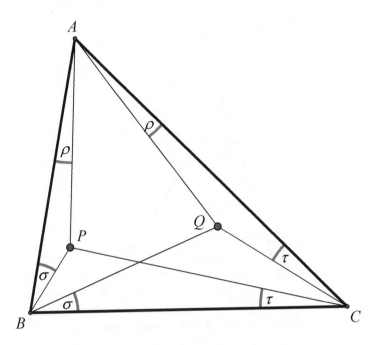

FIGURE 4-4

## THE MIQUEL POINT

The Brocard points were established from concurrent circles. Circles and triangles can also determine other interesting concurrencies. Consider a randomly selected triangle *ABC* and three randomly selected points, *D*, *E*, and *F*, one on each side of the triangle, as shown in figure 4-5. We then draw three circles, where each contains two of the three side points and the included vertex. Lo and behold, the three circles share a common point *M*—the point of concurrency. This point is known as the *Miquel point*, named after the nineteenth-century French mathematician Auguste Miquel, who published this theorem in *Liouville's Journal* in 1838.[1] Yet, as is so often the case in mathematics, there is strong evidence that others had already known about this fantastic relationship—such as the Scottish

mathematician William Wallace (1768–1843) and the Swiss mathematician
Jakob Steiner (1797–1863).

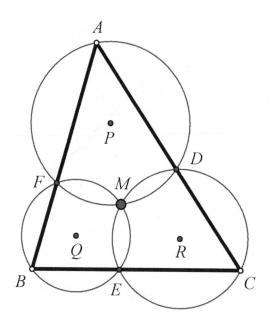

FIGURE 4-5

This concurrency can be easily justified if we recall that the opposite
angles of a quadrilateral inscribed in a circle (called a *cyclic quadrilateral*)
are supplementary. Consider the case when *M* is inside triangle *ABC*, as
shown in figure 4-6. (Bear in mind that a similar argument can be made
for a point *M* outside the triangle.) Points *D*, *E*, and *F* are any points on
sides *AC*, *BC*, and *AB*, respectively, of triangle *ABC*. Let circles *Q* and *R*,
determined by points *F*, *B*, and *E*; and *D*, *C*, and *E*, respectively, meet at *M*.

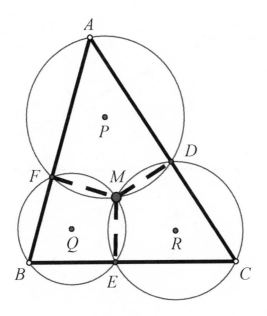

FIGURE 4-6

To justify this concurrency we draw in figure 4-6 *FM*, *EM*, and *DM*. In cyclic quadrilateral *BFME*, we have ∠*FME* = 180° – ∠*B*, since opposite angles of a cyclic quadrilateral are supplementary—that is, they have a sum of 180°. Similarly, in cyclic quadrilateral *CDME*, ∠*DME* = 180° – ∠*C*. By addition, ∠*FME* + ∠*DME* = 360° – (∠*B* + ∠*C*). Therefore, ∠*FMD* = ∠*B* + ∠*C*. However, in triangle *ABC*, ∠*B* + ∠*C* = 180° – ∠*A*. Thus, ∠*FMD* = 180° – ∠*A* and quadrilateral *AFMD* is cyclic. Thus, point *M* lies on all three circles.

In figure 4-7, we see that Miquel's theorem also holds true when the Miquel point, *M*, is outside the original triangle *ABC*.

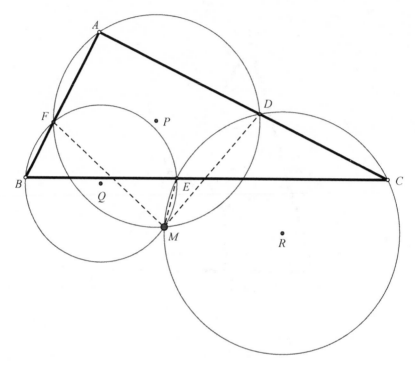

FIGURE 4-7

## THE MIQUEL TRIANGLE

Were that all there was to say about the concurrency of these three circles, we would already have a beautiful phenomenon. However, as you might have expected, there are more gems to be found in this configuration. In figure 4-8, triangle *DEF* is called the *Miquel triangle*. Consider the following: The segments joining the Miquel point of a triangle to the vertices of the Miquel triangle form equal angles with the respective sides of the original triangle. This means that in figure 4-8, $\angle AFM = \angle CDM = \angle BEM (= \mu)$, as well as $\angle ADM = \angle CEM = \angle BFM (= 180° - \mu)$.

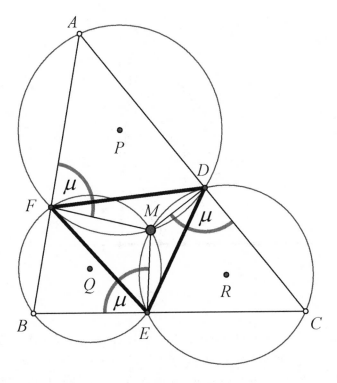

**FIGURE 4-8**

This can be easily justified. Because quadrilateral *AFMD* is cyclic (see figure 4-8), ∠*AFM* is supplementary to ∠*ADM*. But ∠*ADM* is supplementary to ∠*CDM*. Therefore ∠*AFM* = ∠*CDM*, whereupon it follows that ∠*BFM* = ∠*ADM*. The remaining equalities are arrived at by applying the same argument to the other cyclic quadrilaterals.

We say that a triangle is inscribed in a second triangle if each of the vertices of the first triangle lies on the sides of the second triangle. We can then state the following further application of the Miquel point *M*: Two triangles inscribed in the same triangle and having a common Miquel point are similar. We can justify this in the following manner.

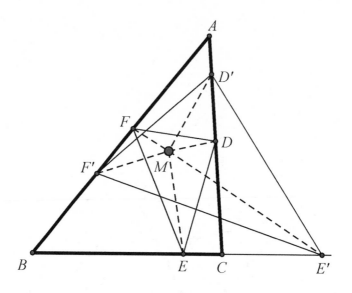

FIGURE 4-9

In figure 4-9, we show triangle $DEF$ and triangle $D'E'F'$, which have the same Miquel point $M$. We have just established that $\angle MFB = \angle MDA$, and $\angle MF'A = \angle MD'C$. Therefore, we have $\Delta MF'F \sim \Delta MD'D$, and $\Delta MD'D \sim \Delta ME'E$. It follows that $\angle FMF' = \angle DMD' = \angle EME'$. By addition, we get $\angle F'MD' \cong \angle FMD$, $\angle F'ME' = \angle FME$, and $\angle E'MD' = \angle EMD$.

Also as a result of the above similar triangles, we get $\frac{MF}{MF'} = \frac{MD}{MD'} = \frac{ME}{ME'}$.

We can establish two triangles similar if two pairs of corresponding sides are proportional and the included angles congruent. This gives us the following pairs of similar triangles:

$\Delta F'MD \sim \Delta FMD$,
$\Delta F'ME' \sim \Delta FME$, and
$\Delta E'MD' \sim \Delta EMD$.

Therefore, $\frac{F'D'}{FD} = \frac{F'M}{FM}$, and $\frac{F'E'}{FE} = \frac{F'M}{FM}$. Thus, $\frac{F'D'}{FD} = \frac{F'E'}{FE}$. Similarly, $\frac{E'D'}{ED} = \frac{F'E'}{FE}$.

This proves that $\Delta DEF \sim \Delta D'E'F'$, since the pairs of corresponding sides are proportional.

We can still show another nifty relationship emanating from the Miquel triangles. That is, the centers of Miquel circles of a given triangle determine a triangle similar to the given triangle. Again, the justification is rather straightforward.

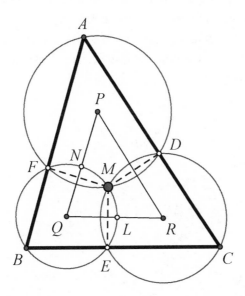

FIGURE 4-10

We can begin to justify this relationship by drawing common chords $FM$, $EM$, and $DM$, as shown in figure 4-10. Let $PQ$ meet circle $Q$ at point $N$ and $RQ$ meet circle $Q$ at point $L$. Since the line of centers of two circles is the perpendicular bisector of their common chord, $PQ$ is the perpendicular bisector of $FM$, so that $\overarc{FN} = \overarc{NM}$. Similarly, $QR$ bisects $\overarc{EM}$ so that $\overarc{ML} = \overarc{LE}$.

Now, the central angle $\angle NQL = (\overarc{NM} + \overarc{ML}) = \frac{1}{2}(\overarc{FE})$ and the inscribed angle $\angle FBE = \frac{1}{2}(\overarc{FE})$. Therefore, $\angle NQL = \angle FBE$.

In a similar fashion it may be proved that $\angle QPR = \angle BAC$. Thus, $\triangle PQR \sim \triangle ABC$, since their corresponding angles are equal.

You might wish to investigate the Miquel triangle of an equilateral triangle, or the Miquel triangle of a right triangle, as they yield some interesting properties.

## CONCYCLIC POINTS

Auguste Miquel is also responsible for another truly amazing relationship that also involves intersections of circles. We begin with a randomly drawn (irregular) pentagram (a five-cornered star) as shown in figure 4-11. We will then draw the circumscribed circle for each of the five outside triangles (shaded). Amazingly, the intersection points ($A$, $B$, $C$, $D$, and $E$) of the five circles always lie on the same circle (dashed line).[2] Such points are said to be *concyclic*.

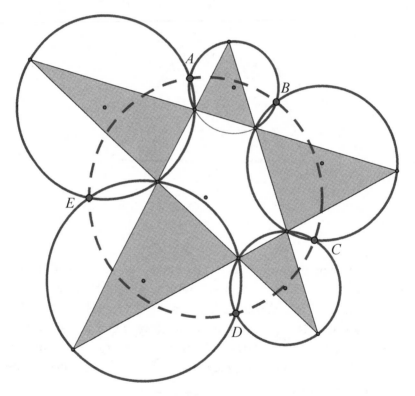

FIGURE 4-11

There is a somewhat similar "five-circle theorem" that states that if five consecutively intersecting circles, whose centers lie on a circle, are drawn so that one of their two intersection points with their adjacent circle also lies on the circle of centers, then by joining the remaining intersection points a (not necessarily regular) pentagram *PQRST* is formed with its vertices each lying on one of the circles. This is shown in figure 4-12.

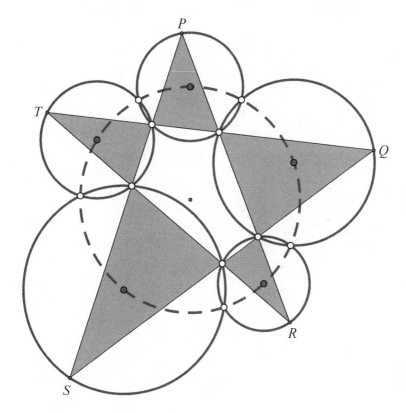

FIGURE 4-12

## MORE CONCURRENT CIRCLES

Before we leave the topic of concurrency of circles, we should admire another configuration that opens the door to lots of further such investigations. In figure 4-13, we have a randomly drawn triangle *ABC* and on each of its sides we have constructed the reflection of triangle *ABC*. Consequently, all four triangles are congruent ($\triangle ABC \cong \triangle A'BC \cong \triangle AB'C$

$\cong \Delta ABC'$). First, we notice that the three circumscribed circles of the reflected triangles are concurrent at the point $Y$. Then when we draw lines joining each of the centers of these circles with the remote vertex of the original triangle $ABC$, we find that these lines are also concurrent at a point $X$.

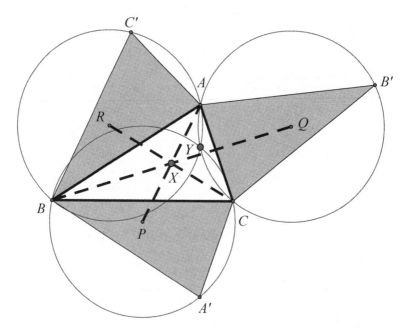

FIGURE 4-13

The beauty here is that the configurations that generate concurrency are limited only by one's imagination. Suppose we take the original triangle $ABC$ and rotate it 180° about the midpoint $(M_a, M_b, M_c)$ of each of its sides. We will then get the darker-shaded triangles: $\Delta A''BC$, $\Delta AB''C$, and $\Delta ABC''$, shown in figure 4-14. Once again, the circumscribed circles of each of the rotated triangles (shaded) are concurrent at point $Y$.

And furthermore, also the lines joining each of the centers, $P$, $Q$, and $R$, of these circles with the vertices $A''$, $B''$, and $C''$, respectively, of the rotated triangle $ABC$, are also concurrent at this point $Y$.

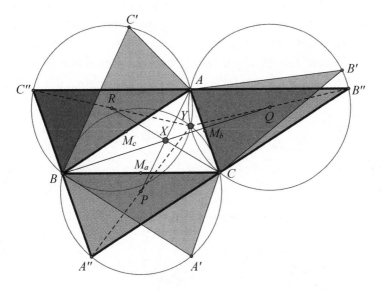

FIGURE 4-14

In addition, as in the previous example, the lines joining each of the centers of these circles with the remote vertex of the original triangle *ABC* are also concurrent at a point *X*. We leave other such concurrency findings to the reader.

To close out this chapter on concurrent circles there is a cute series of concurrent circles and points first published by the American mathematician Roger A. Johnson (1890–1954),[3] where for a given triangle three concurrent congruent circles are drawn (these are often called *Johnson circles*), with each containing two vertices of the triangle. In figure 4-15, we have three concurrent congruent circles with centers $J_a$, $J_b$, and $J_c$, with each containing two vertices of the triangle *ABC*. The following surprising properties hold true:

- The three concurrent circles are unique and are congruent.
- The three concurrent Johnson circles have the same radius as the circumscribed circle of the original triangle *ABC*, and their centers, $J_a$, $J_b$, and $J_c$, lie on a circle with the same radius as that of the three concurrent circles.

- The point of intersection of the three concurrent *Johnson circles* is the orthocenter *H* (point of intersection of the altitudes) of the original triangle *ABC*.
- The triangle formed by the centers of the concurrent circles, triangle $J_a J_b J_c$ (*Johnson triangle*) is congruent to the original triangle *ABC*.

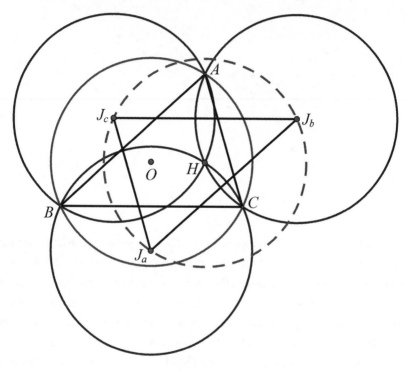

FIGURE 4-15

With this amazing relationship, we have a fine way to close our chapter on concurrent circles.

# CHAPTER 5

# SPECIAL LINES OF A TRIANGLE

I n this chapter, we will revisit the special lines of a triangle. In so doing, we will reveal many surprising relationships that further demonstrate how the triangle harbors many secrets that are constantly being discovered by mathematicians and amateurs who love to delve more deeply into Euclidean geometry with the goal of stumbling onto sometimes-unexpected properties of the triangle and its many parts.

## THE ANGLE BISECTORS OF A TRIANGLE

We touched on a most important property of the angle bisectors of a triangle when we showed earlier that they are concurrent (p. 47). Now let's inspect a few other properties of the angle bisectors of a triangle. For example, two angle bisectors in a triangle, as shown in figure 5-1, intersect to form an angle that is equal to a right angle plus one-half of the third angle of the triangle.

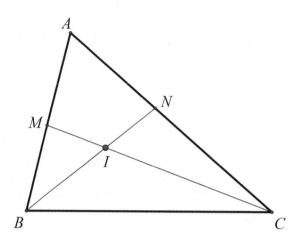

FIGURE 5-1

In figure 5-1, we have triangle angle bisectors $BN$ and $CM$ intersecting at point $I$. It is rather easy to justify that $\angle BIC = 90° + \frac{1}{2} \angle A$. We begin by considering triangle $BIC$. $\angle BIC = 180° - \angle IBC - \angle ICB$.

Then $\angle BIC = 180° - \frac{1}{2} \angle ABC - \frac{1}{2} \angle ACB$.

However, $\angle ABC + \angle ACB = 180° - \angle A$. It then follows that $\frac{1}{2} \angle ABC + \frac{1}{2} \angle ACB = 90° - \frac{1}{2} \angle A$.

By substituting this value into the previous equation, we get our desired result. First,

$\angle BIC = 180° - \left(90° - \frac{1}{2}\angle A\right)$, from which we get $\angle BIC = 90° + \frac{1}{2} \angle A$.

This relationship can be extended to the *exterior* angle bisectors of a triangle, where one of the angles formed is equal to a right angle *minus* one-half of the third angle of the triangle. In figure 5-2, we have exterior angle bisectors $BJ$ and $CJ$ meeting at point $J$. We will use a technique similar to that used above to establish this relationship.

We begin with $\angle BJC = 180° - \frac{1}{2} \angle EBC - \frac{1}{2} \angle FBC$

$= 180° - \frac{1}{2} (180° - \angle ABC) - \frac{1}{2} (180° - \angle ABC)$

$= 180° - 90° + \frac{1}{2} \angle ABC - 90° + \frac{1}{2} \angle ABC$

$= \frac{1}{2} (\angle ABC + \angle ABC) = \frac{1}{2} (180° - \angle A)$.

Therefore, $\angle BJC = 90° - \frac{1}{2} \angle A$.

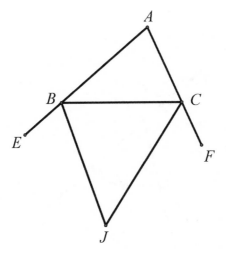

FIGURE 5-2

The angle bisector not only bisects the angle, it also partitions the opposite side proportionally to the two adjacent sides. That is, for figure 5-3, we have $AD$ as the bisector of angle $BAC$, which then gives us $\dfrac{c}{b} = \dfrac{m}{n}$.

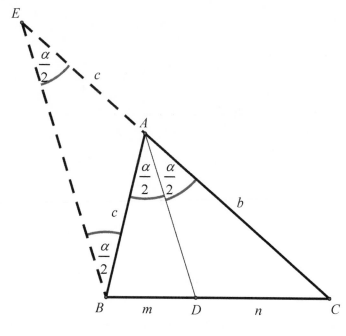

FIGURE 5-3

To show why this is true is a bit tricky, as we must bring in some auxiliary lines: We construct $BE$ parallel to $AD$ and intersecting $CA$ (extended) at point $E$. All the angles marked with $\frac{\alpha}{2}$ are equal. Therefore, since triangle $AEB$ is isosceles, $AE = AB = c$. For triangle $BCE$, we have $\frac{AE}{CA} = \frac{BD}{CD}$, or $\frac{c}{b} = \frac{m}{n}$.

While it is relatively easy to find the length of an altitude of a triangle (usually using the Pythagorean theorem), it is not so trivial to find the length of an angle bisector of a triangle. There is a very helpful relationship that can be used to find the length of an angle bisector.

In figure 5-4, we have triangle $ABC$, with $AD$, the bisector of $\angle BAC$, which we will call $t_a$. We begin our development by extending $AD$ beyond $D$ to meet the circumscribed circle of triangle $ABC$ at $E$. We then draw $BE$. Since $\angle BAD = \angle CAD$, and $\angle E = \angle C$ (both angles inscribed in the same arc),

$$\triangle ABE \sim \triangle ADC, \text{ and } \frac{AC}{AD} = \frac{AE}{AB},$$

or $AC \cdot AB = AD \cdot AE = AD (AD + DE) = AD^2 + AD \cdot DE.$          (I)

However, $AD \cdot DE = BD \cdot DC,$                                    (II)
since they are intersecting chords of the same circle.

Substituting equation (II) into equation (I), we obtain $AD^2 = AC \cdot AB - BD \cdot DC$, which is what we sought to develop. Yet in simpler terms, using the letter designations in figure 5-3, we can write this relationship as $t_a^2 = bc - m \cdot n$, or $t_a = \sqrt{bc - mn}$ .

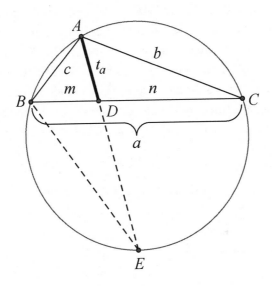

FIGURE 5-4

For the ambitious reader we provide yet another formula for finding the length of the angle bisector of a triangle given only the three sides:

$$t_a = \frac{2 \cdot \sqrt{bcs(s-a)}}{b+c}, t_b = \frac{2 \cdot \sqrt{cas(s-b)}}{c+a}, t_c = \frac{2 \cdot \sqrt{abs(s-c)}}{a+b}, \text{ where } s = \frac{a+b+c}{2}.$$

## THE SYMMEDIANS OF A TRIANGLE

As we did earlier, we will once again construct a square on each side of our randomly drawn triangle $ABC$ as shown in figure 5-5. However, this time we will construct a triangle by extending the remote sides of each of the squares—giving us triangle $A'B'C'$. When we draw the lines joining each vertex of the newly formed triangle $A'B'C'$ to the nearest vertex of the original triangle ($ABC$), we find that these three lines are concurrent at point $K$.

However, these three lines, $A'AK$, $B'BK$, and $C'CK$, and their point of intersection, $K$, have other special properties. To see one of these properties, we first draw the three medians, $A'P$, $B'Q$, and $C'R$ (meeting at point $G$), of triangle $A'B'C'$. The angle formed by each of the three lines, $A'AK$, $B'BK$, and $C'CK$, with a side of each of the angles of triangle $A'B'C'$ is equal to the angle formed by the median and the other side of each of the angles of the triangle. That is, $\angle B'A'K = \angle C'A'G$, $\angle A'B'K = \angle C'B'G$, and $\angle B'C'K = \angle A'C'G$.

Hence, these three lines, $A'AK$, $B'BK$, and $C'CK$, are called *symmedians*, as they are, in a sense, symmetric to the medians of the triangle.

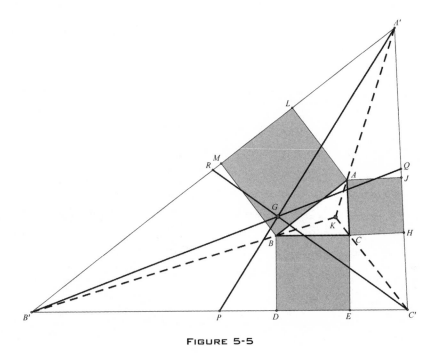

FIGURE 5-5

These lines, the symmedians, and their point, $K$, of concurrency have many interesting properties. We have, for example, an unexpected proportionality: The distance from the point of concurrency, also called the *symmedian point*,[1] to the sides is proportional to the respective sides. That is,

$$\frac{KW}{KV} = \frac{A'B'}{A'C'}, \frac{KV}{KU} = \frac{A'C'}{B'C'}, \text{ and } \frac{KW}{KU} = \frac{A'B'}{B'C'}. \text{ (See figure 5-6.)}$$

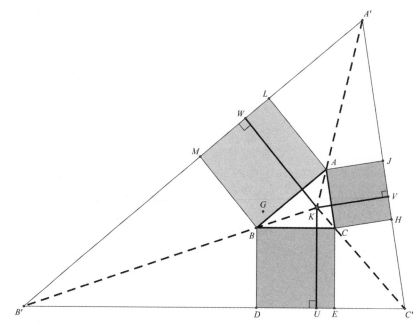

FIGURE 5-6

We get the inverse proportion when we consider the distances from the midpoint ($P$) of a side ($B'C'$) of the triangle to the other two sides. Since the median of a triangle partitions the triangle into two equal areas, as shown in figure 5-7, *Area $\triangle A'PB' = $ Area $\triangle A'PC'$*, and $\frac{1}{2} A'B' \cdot PR = A'C' \cdot PQ$.

Therefore, $\frac{PR}{PQ} = \frac{A'C'}{A'B'}$, which is the inverse relationship from that determined by the symmedian point.

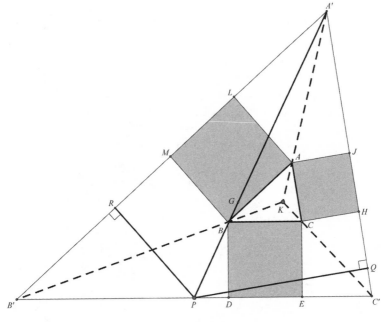

FIGURE 5-7

We should note that the symmedian point (point $K$ in figure 5-6) of a triangle ($A'B'C'$) has some very interesting properties, one of which is that the sum of the squares of the distances from the symmedian point to the three sides of the triangle is a minimum. That is, in figure 5-6, the sum of the squares of the distances from the symmedian point $K$ to the three sides, $KU^2 + KV^2 + KW^2$, has the least value of the sum of the squares of the perpendiculars from any other point in the plane of the triangle to the sides of the triangle.

The symmedian point is also locatable in another way. In figure 5-8, we find that the line joining the midpoint $R$ of the altitude $A'Q$ with the midpoint $P$ of the side $B'C'$ of the triangle $A'B'C'$ to which the altitude is drawn contains the symmedian point $K$. Therefore, if we were to do this for the other altitudes of the triangle, we would find the common point of intersection—namely, the symmedian point.

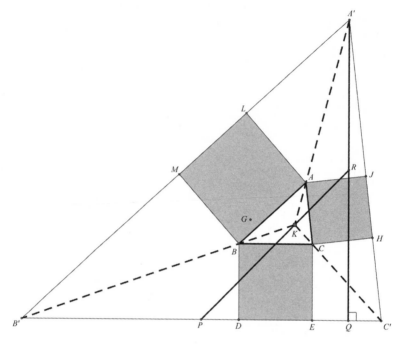

FIGURE 5-8

Having previously shown that the angle bisector partitions the side to which it is drawn proportional to the adjacent sides (p. 47), in figure 5-9, we now have another point $(K_a)$ that will determine a proportionality to the sides of a triangle. That point $(K_a)$ is the intersection of a symmedian with a side $(a)$ of the triangle. From this point the distances to the two adjacent sides are proportional to these two sides. In figure 5-9 we, therefore, have $\dfrac{K_aD}{K_aE} = \dfrac{AB}{AC}$.

The square of this ratio, $\left(\dfrac{AB}{AC}\right)^2$, is also related to the symmedian of a triangle. The symmedian divides the side of a triangle to which it is drawn in the ratio of the *squares* of the remaining two sides. Consider the symmedian $AK_a$ of triangle $ABC$ (figure 5-9). We can easily show that $\dfrac{BK_a}{K_aC} = \left(\dfrac{AB}{AC}\right)^2$.

To begin this justification, we note that the ratio of the areas of the two triangles into which the symmedian partitioned the original triangle is equal to the ratio of the segments determined along the base of the triangle, because they share that same altitude, $AH_a$.

Therefore,

$$\frac{BK_a}{K_aC} = \frac{Area\ \Delta ABK_a}{Area\ \Delta AK_aC} = \frac{\frac{1}{2}\cdot AB\cdot K_aD}{\frac{1}{2}\cdot AC\cdot K_aE} = \frac{AB}{AC}\cdot\frac{K_aD}{K_aE}.$$

We can then conclude that $\dfrac{BK_a}{K_aC} = \left(\dfrac{AB}{AC}\right)\cdot\left(\dfrac{AB}{AC}\right) = \left(\dfrac{AB}{AC}\right)^2.$

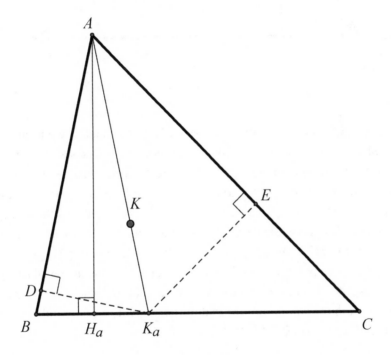

FIGURE 5-9

We can also relate the symmedian to an altitude and a median of a right triangle. In figure 5-10, consider the altitude $AH_a$ to the hypotenuse $BC$ of right triangle $ABC$. The line joining the midpoint, $M$, of altitude $AH_a$ to vertex $B$ is a symmedian $(BK_b)$ of the triangle. Remember, to be a symmedian, $BMK_b$, must be such that $\angle ABM = \angle ABK_b = \angle CBM_b$, which is easily proved by showing that $\Delta ABH_a \sim \Delta CBA$, since they are right triangles that share $\angle ABC$. Then the median to the corresponding legs of these similar triangles must form equal angles, respectively.

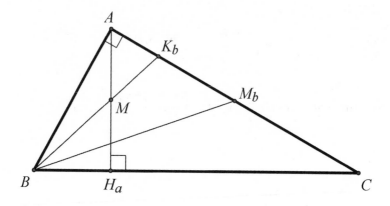

FIGURE 5-10

Points of concurrency in a triangle are very interrelated to one another, as we have consistently seen throughout this chapter. One such example is that the symmedian point of a triangle is the centroid of its *pedal triangle*. The pedal triangle of a given triangle is formed by joining the feet of the perpendiculars from a given point to each of the three sides of the triangle. (The foot of a perpendicular is the point at which the perpendicular line intersects the line to which it is perpendicular.) In figure 5-11, point $K$ is the symmedian point of triangle $ABC$. Triangle $A'B'C'$ is the pedal triangle with respect to point $K$. Since the medians of triangle $A'B'C'$ contain point $K$, it is the centroid of the pedal triangle, where $D$, $E$, and $F$ are the midpoints of the sides $B'C'$, $A'C'$, and $A'B'$.

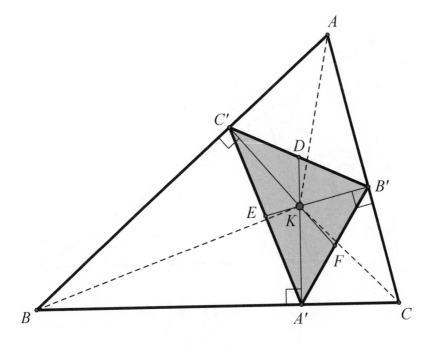

FIGURE 5-11

Recall the Gergonne point of a triangle (see p. 50 and figure 2-15). In figure 5-12, we have the lines $AP_a$, $BP_b$, and $CP_c$, which determine the Gergonne point[2] $G$ of triangle $ABC$ and are the symmedians of the Gergonne triangle $P_a P_b P_c$. We leave this to the reader to justify.

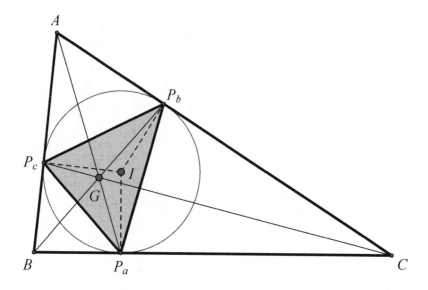

FIGURE 5-12

## THE ALTITUDES OF A TRIANGLE

We showed earlier that the altitudes of a triangle are concurrent and that the point of concurrency is called the orthocenter of the triangle. When we connect the feet of the altitudes, we form a pedal triangle of the original triangle. The amazing result is that the altitudes of the original triangle are the angle bisectors of the pedal triangle. This can be seen in figures 5-13a and 5-13b, where the pedal triangle $H_aH_bH_c$ has the altitudes, $AH_a$, $BH_b$, and $CH_c$ as its angle bisectors.

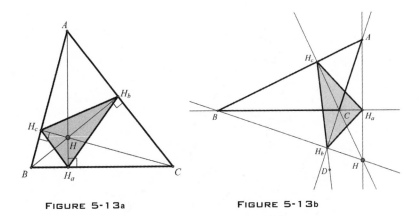

FIGURE 5-13a          FIGURE 5-13b

Furthermore, this pedal triangle—formed by the feet $H_a$, $H_b$, $H_c$ of the altitudes—partitions the original triangle into three triangles, each of which is similar to the original one. That is, for figure 5-13,

$$\triangle ABC \sim \triangle AH_bH_c \sim \triangle H_aBH_c \sim \triangle H_aH_bC.$$

This type of pedal triangle has the unique feature of having the least perimeter of all the triangles that are placed with one vertex on each side of the original acute-angled triangle. For example, in figure 5-14, we place another "inscribed triangle" (*XYZ*) in triangle *ABC*. The perimeter of triangle *XYZ* is greater than the perimeter of triangle $H_aH_bH_c$.

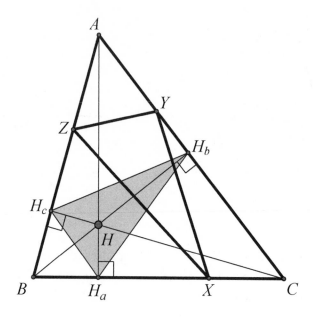

FIGURE 5-14

We will now extend the three altitudes of a triangle through the side to which they are perpendicular to intersect the triangle's circumscribed circle. The unexpected result is that the arcs determined by these extended altitude intersections with the circumscribed circle are bisected by the vertices of the original triangle. This can be seen in figure 5-15, where $P$, $Q$, and $R$ are the points at which the altitude extended through its base intersects the circumscribed circle of triangle $ABC$. These points determine arcs $\overset{\frown}{PQ}$, $\overset{\frown}{RQ}$, and $\overset{\frown}{PR}$, whose midpoints are the vertices of the original triangle, namely, $C$, $A$, and $B$, respectively.

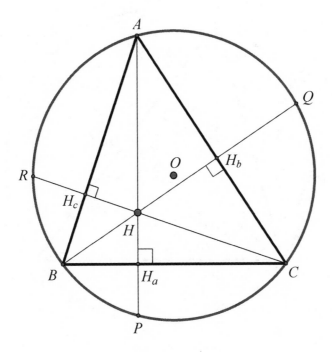

FIGURE 5-15

If we now connect points $P$, $Q$, and $R$ to get triangle $PQR$, we can see that $\triangle PQR \sim \triangle H_a H_b H_c$, with their corresponding sides mutually parallel. (See figure 5-16.)

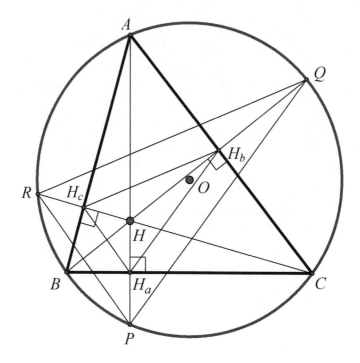

FIGURE 5-16

We can find lots of relationships with the pedal triangle, such as when we consider the tangency points $X$, $Y$, and $Z$ of its inscribed circle, as shown in figure 5-17. If we extend line $XY$ to meet $AH_a$ at $P$, and extend $YZ$ to meet $CH_c$ at $R$, then we have created a parallelogram $PH_bRY$. Furthermore, we can show that $PY = PZ$, and that $PY$ is parallel to $AB$ and $RY$ is parallel to $BC$.

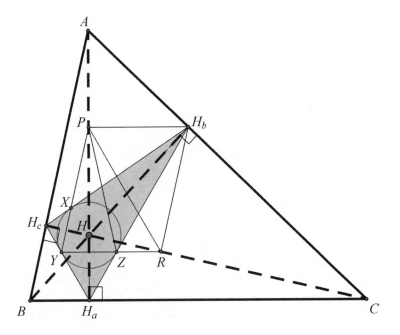

FIGURE 5-17

Moreover, this configuration also has six concyclic points: $P, H_b, R, Z,$ $H$, and $X$, as shown in figure 5-18.

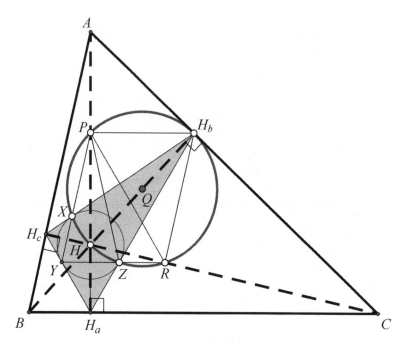

FIGURE 5-18

Although we will revisit the pedal triangle later in this chapter, there are still many more properties in this configuration that can be fun to discover, such as the relationship of the circumscribed circle's radius to the pedal triangle. We leave these for the reader to discover.

## GENERAL CEVIANS

Finding the length of just any cevian, not necessarily an angle bisector or an altitude of a triangle, was for a long time thought impossible. Yet the problem was first solved by the famous Scottish geometer Robert Simson (1687–1768), who presented it in lectures, but allowed his notes to be used by his prize student, Matthew Stewart (1717–1785), in his famous publication, *General Theorems of Considerable Use in the Higher Parts of Mathematics*

(Edinburgh, 1746). Simson's generosity was motivated by his desire to see Stewart obtain the chair of Mathematics at the University of Edinburgh. He was successful. It is interesting to note how Simson was credited with a theorem he did not know (pp. 136, 137), yet not credited with one that he deserved to have to his credit. We shall still refer to the theorem by the author, Stewart, of the book in which it first appeared in print.

Actually, Simson deserves particular note for his definitive book, *The Elements of Euclid* (Glasgow, 1756), which remained in print for over 150 years. This book is the basis for all subsequent study of Euclid's *Elements*, including the high-school geometry courses taught in the United States today.

Although Stewart's theorem will appear quite cumbersome, it is extremely powerful in that it allows us to find the length of any cevian of a triangle, not just the more common ones, such as angle bisectors, medians, or altitudes. Using the letter designations in figure 5-19, Stewart's theorem states the following relationship for all kinds of triangles: $b^2m + c^2n = a(d^2 + mn)$, where $d = AD$ is any cevian of the triangle $ABC$. Even though it is more complex than the formula we previously developed for an angle bisector, it is worth knowing, since its applications are much more extensive. The derivation is provided in the appendix.

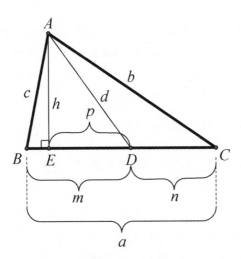

FIGURE 5-19

Stewart's theorem leads to many nice triangle relationships. Consider the following relationship that we will develop by using this theorem: In the right triangle $ABC$ shown in 5-20, the sum of the squares of the distances $p$ and $q$ from the vertex $A$ of the right angle to the trisection points $D$ and $E$ along the hypotenuse is equal to $\frac{5}{9}$ the square of the measure of the hypotenuse $a$. In figure 5-20, this would be $p^2 + q^2 = \frac{5}{9} a^2$.

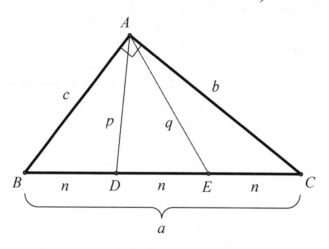

FIGURE 5-20

To justify this relationship we first apply Stewart's theorem twice to figure 5-20. First, using $p$ as the internal line segment, we find that

$$b^2n + 2c^2n = a(p^2 + 2n^2), \tag{I}$$

where $p = AD$ is a cevian of the triangle $ABC$.

Then using $q$ as the internal line segment we get

$$2b^2n + c^2n = a(q^2 + 2n^2). \tag{II}$$

Adding equations (I) and (II) results in $3n(b^2 + c^2) = a(p^2 + q^2 + 4n^2)$. Applying the Pythagorean theorem to triangle $ABC$, we get $b^2 + c^2 = a^2$, and then substituting this in the above equation, we find that $3na^2 = a(p^2 + q^2 + 4n^2)$.

Since $3n = a$, we get $a^2 = p^2 + q^2 + 4n^2$.
But we have $2n = \dfrac{2}{3} a$; therefore,

$$p^2 + q^2 = a^2 - 4n^2 = a^2 - (2n)^2 = a^2 - \frac{4}{9}a^2 = \frac{5}{9} a^2.$$

Simply stated, our result is what we sought to show: $p^2 + q^2 = \dfrac{5}{9}a^2$.

## THE MEDIANS OF A TRIANGLE REVISITED

We know that the medians of a triangle connect a vertex of a triangle with the midpoint of the opposite side. We also established earlier that the medians of a triangle are concurrent, as a matter of fact, this point of concurrency is the center of gravity,[3] or *centroid*, of the triangle, and is located two-thirds the distance from the vertex to the opposite side. Some lesser-known properties of the medians exhibit some of the hidden treasures they hold.

Some nice relationships involving the medians of a triangle involve the *squares* of the lengths of the medians of a given triangle. In order to do that, we will first reach back to Stewart's theorem and apply it to triangle *ABC*, with median $AM_a$. (See figure 5-21.)

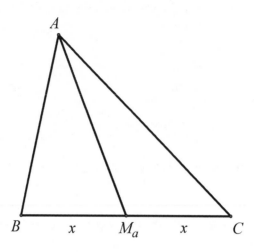

FIGURE 5-21

First, we will show that twice the square of the length of a median of a triangle equals the sum of the squares of the lengths of the two including sides minus one-half the square of the length of the third side. Although this may sound a bit complicated, it will help us find a relationship for the sum of the squares of the medians of the triangle. Follow along as we use Stewart's theorem and then do some basic algebraic simplifications.

By applying Stewart's theorem to triangle $ABC$, we get the following:

$$AC^2 \cdot BM_a + AB^2 \cdot M_aC = (BM_a + M_aC) \cdot (AM_a^2 + BM_a \cdot M_aC).$$

To simplify the discussion we will let $x = BM_a = M_aC \ (= \frac{a}{2})$, which then allows us to simplify this complicated equation. We thus can restate the equation in a more manageable fashion as

$$AC^2 \cdot x + AB^2 \cdot x = (x + x) \cdot (AM_a^2 + x^2) \qquad \text{I divide both sides by } x$$
$$AC^2 + AB^2 = 2(AM_a^2 + x^2)$$
$$2\,AM_a^2 = AB^2 + AC^2 - 2x^2$$

Since $x = \dfrac{BC}{2} = \dfrac{a}{2}$, we obtain our desired result:

$$2AM_a^2 = AB^2 + AC^2 - \frac{BC^2}{2}, \text{ or } 2\,m_a^2 = b^2 + c^2 - \frac{a^2}{2}.$$

Analogously, we get $2\,m_b^2 = a^2 + c^2 - \dfrac{b^2}{2}$, and $2\,m_c^2 = a^2 + b^2 - \dfrac{c^2}{2}$.

You may find this relationship a bit cumbersome, yet it helps us to prove some rather useful and interesting properties of medians of a triangle, such as the following:

The sum of the squares of the lengths of the medians of a triangle equals three-fourths the sum of the squares of the lengths of the sides of the triangle.

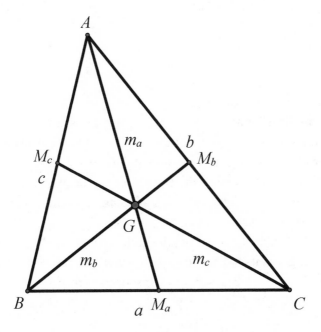

FIGURE 5-22

Using our newly developed relationship (above) and applying it to each of the three medians of triangle *ABC* (figure 5-22), we get

$$2m_a^2 = b^2 + c^2 - \frac{1}{2}a^2$$

$$2m_b^2 = a^2 + c^2 - \frac{1}{2}b^2$$

$$2m_c^2 = a^2 + b^2 - \frac{1}{2}c^2$$

By adding these three equations, we find that

$$2\left(m_a^2 + m_b^2 + m_c^2\right) = 2\left(a^2 + b^2 + c^2\right) - \frac{1}{2}\left(a^2 + b^2 + c^2\right),$$

$$2\left(m_a^2 + m_b^2 + m_c^2\right) = \frac{3}{2}\left(a^2 + b^2 + c^2\right),$$

$$m_a^2 + m_b^2 + m_c^2 = \frac{3}{4}\left(a^2 + b^2 + c^2\right),$$

which was our desired result.

We may then immediately use this result to establish a relationship between the sum of the squares of the lengths of the segments joining the centroid with the vertices, and the sum of the squares of the lengths of the sides as follows: The sum of the squares of the lengths of the segments joining the centroid with the vertices is one-third the sum of the squares of the lengths of the sides.

Since the length of a segment joining the centroid with a vertex is two-thirds the length of its respective median, we can represent these segments as $\frac{2}{3}m_a, \frac{2}{3}m_b,$ and $\frac{2}{3}m_c$. The sum of their squares is then

$$\left(\frac{2}{3}m_a\right)^2 + \left(\frac{2}{3}m_b\right)^2 + \left(\frac{2}{3}m_c\right)^2 = \frac{4}{9}\left(m_a^2 + m_b^2 + m_c^2\right).$$

However, from the previously developed relationship, we have $m_a^2 + m_b^2 + m_c^2 = \frac{3}{4}\left(a^2 + b^2 + c^2\right).$

Therefore by substitution we find that

$$\frac{4}{9}\left(m_a^2 + m_b^2 + m_c^2\right) = \frac{4}{9} \cdot \frac{3}{4}\left(a^2 + b^2 + c^2\right) = \frac{1}{3}\left(a^2 + b^2 + c^2\right),$$

which is what was to be demonstrated.

We can now take this a step further by showing how the sum of the squares of the distances from *any* point in a triangle relates to the vertices of the triangle. Here we will have another of the hidden gems of triangles.

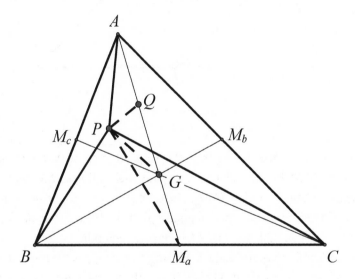

FIGURE 5-23

In figure 5-23 we let point *P* be *any* point in triangle *ABC* with centroid *G*. The following is then true: $AP^2 + BP^2 + CP^2 = AG^2 + BG^2 + CG^2 + 3PG^2$. We provide a proof of this curious result in the appendix.

The medians of a triangle provide us with many interesting relationships, such as in any triangle, a median and the *midline* that intersects it (in the interior of the triangle) bisect each other. (A midline of a triangle is the line joining the midpoints of two sides of the triangle.)

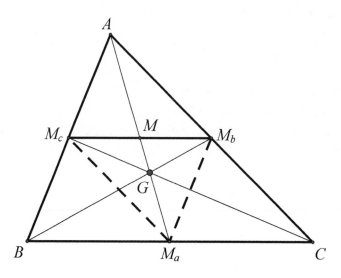

FIGURE 5-24

To show this is true for figure 5-24 we need to show that median $AM_a$ and midline $M_bM_c$ bisect each other. By drawing midlines $M_bM_c$ and $M_aM_b$, we form parallelogram $AM_cM_aM_b$, since both pairs of opposite sides are parallel. Therefore, since the diagonals of a parallelogram bisect each other, we can state that $AM_a$ and $M_bM_c$ bisect each other.

As we mentioned earlier, the centroid of a triangle is the point at which the triangle can be balanced. That is, if we take a cardboard triangle, it will balance on a pin located at the centroid. This balancing characteristic of the centroid may be seen also in another rather unusual way. Consider the triangle $ABC$ in figure 5-25, and let $XYZ$ be any line through the centroid $G$ and separating vertices $B$ and $C$ from vertex $A$. When perpendiculars are drawn from each vertex of triangle $ABC$ to this line $XYZ$, we find that $AX = BY + CZ$. Remember, the line $XYZ$ was any randomly drawn line through the centroid of the triangle.

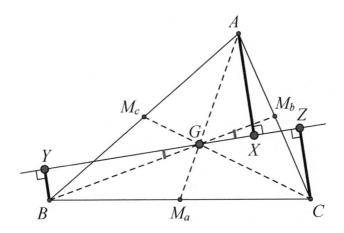

FIGURE 5-25

With dynamic geometry software, such as the Geometer's Sketchpad®
or GeoGebra®, you can rotate the line $XYZ$ about point $G$ and find the
relationship $AX = BY + CZ$ will always hold true when the line through the
centroid $G$ separates vertices $B$ and $C$ from vertex $A$—quite an amazing
aspect of the centroid's property. A justification of this relationship can be
found in the appendix.

We have just taken a small sample of the many lines that expose some
truly delightfully interesting relationships with triangles. There are many
more that we will leave to the reader to discover!

# CHAPTER 6

# USEFUL TRIANGLE THEOREMS

Although we have encountered many surprising relationships throughout the first five chapters, we find that there are some theorems that are particularly noteworthy, not only because they make our investigation of triangle properties much more efficient, but also because they expose many additional triangle properties. Each of these seems to provide a very surprising relationship in triangles and should serve as a springboard for further investigations.

## MENELAUS'S THEOREM

Earlier we encountered the powerful relationship discovered by Giovanni Ceva that gave us concurrency from points on the sides of a triangle that determined equal products of alternate segments. Analogous to that relationship is one that Ceva discovered to have been developed in about 100 CE by Menelaus of Alexandria (70–140 CE), who established that the equal products of alternate segments on the sides of a triangle determine collinear points, as you can see from the following statement of Menelaus's theorem:

> If three points, $X$, $Y$, and $Z$, are located where each of two of these points are on two different sides of triangle $ABC$, and the third point is on the extension of the third side of the triangle such that $AZ \cdot BX \cdot CY = AY \cdot BZ \cdot CX$, then the three points $X$, $Y$, and $Z$ are collinear.

This can be seen in figures 6-1a and 6-1b.

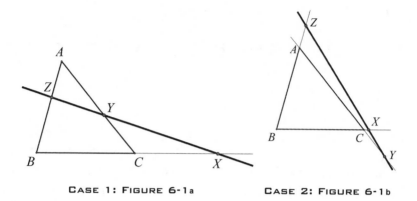

CASE 1: FIGURE 6-1a          CASE 2: FIGURE 6-1b

The proof of this theorem can be found in the appendix. The converse of this theorem is also true. Namely, if the three points are collinear, with one on each side of a triangle (or extension), then the products of the alternate segments of the sides of the triangle are equal. We will use Menelaus's theorem as we further explore lines and points of a triangle—more specifically when we consider points on a triangle that are collinear.

## SIMSON'S THEOREM

When considering collinear points involving triangles, the famous Simson theorem must be acknowledged. One of the great injustices in the history of mathematics involves this theorem. It was originally published by William Wallace (1768–1843) in Thomas Leybourn's *Mathematical Repository* (1799–1800), which through careless misquotes has been attributed to Robert Simson (1687–1768), the famous English interpreter of Euclid's *Elements*, whose book—as we mentioned earlier—has been the basis for the study of geometry in the English-speaking world and more specifically has greatly influenced the American high-school geometry course. To conform to the norm, we shall use the popular reference *Simson's theorem* throughout this book.

Simson's theorem states that the feet of the perpendiculars drawn from *any* point on the circumscribed circle of a triangle to the sides of the triangle are collinear. This is shown in figure 6-2, where point *P* is any point on the circumscribed circle of triangle *ABC*. We then draw $PY \perp AC$ at *Y*, $PZ \perp AB$ at *Z*, and $PX \perp BC$ at *X*. According to Simson's theorem, points *X*, *Y*, and *Z* are collinear. This line is usually referred to as the *Simson line* of *P* with respect to triangle *ABC*. It should be noted that the converse of this theorem is also true.

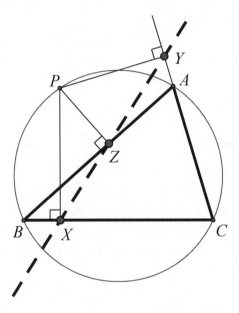

FIGURE 6-2

Before we discuss a number of aspects of this famous line, we will provide a proof of the theorem. Although Menelaus's theorem can be used to prove Simson's theorem (see appendix), we will provide a more elementary proof here.

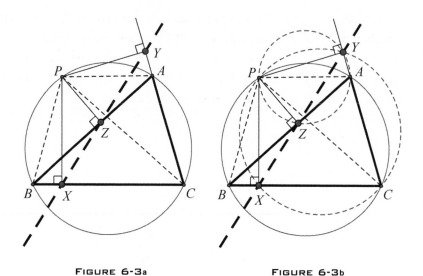

FIGURE 6-3a                    FIGURE 6-3b

In figure 6-3a, we will draw *PA*, *PB*, and *PC*. Since $\angle PYA$ and $\angle PZA$ are right angles, they are supplementary, thus establishing quadrilateral *PZAY* as cyclic (figure 6-3b shows the concyclic points).

(Recall that a *cyclic quadrilateral* is one that can be inscribed in a circle.)

Therefore, $\angle PYZ = \angle PAZ$.                                          (I)

Similarly, since $\angle PYC$ is supplementary to $\angle PXC$, quadrilateral *PXCY* is cyclic, and

$\angle PYX = \angle PCB$.                                                   (II)

However, quadrilateral *PACB* is also cyclic, since it is inscribed in the given circumscribed circle, and therefore

$\angle PAZ = \angle PCB$.                                               (III)

From equations (I), (II), and (III), we can establish that $\angle PYZ = \angle PYX$, and thus points $X$, $Y$, and $Z$ are collinear.

A curiosity of the Simson line can be seen when we generate the Simson line from the intersection point on the circumscribed circle of the extension of one of the triangle's altitudes. This Simson line is parallel to the tangent at the vertex from which this altitude emanates. For example, when, in figure 6-4, the altitude $BD = h_b$ of triangle $ABC$ meets the circumscribed circle at $P$ (and at $B$), then the Simson line of triangle $ABC$ with respect to $P$ is parallel to the line tangent to the circle at $B$.

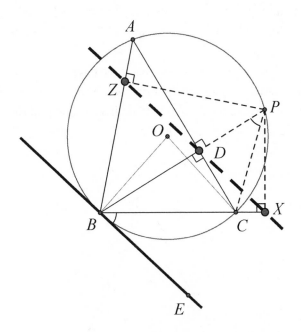

FIGURE 6-4

This can be rather easily justified. We know that in figure 6-4 point $D$ is one of the points on the Simson line. We also have $PX$ and $PZ$ perpendicular, respectively, to sides $BC$ and $AB$ of triangle $ABC$. Therefore, points $X$, $D$, and $Z$ determine the Simson line of $P$ with respect to triangle $ABC$.

Next, we will draw *PC*. Consider quadrilateral *PDCX*, where ∠*PDC* = ∠*PXC* = 90°, thus making *PDCX* a cyclic quadrilateral. The two inscribed angles intercepting the same arc $\overset{\frown}{DC}$ of the circumscribed circle of quadrilateral *PDCX* are equal.

Therefore, ∠*DXC* = ∠*DPC*.                                                       (I)

However, in the circumscribed circle (with circumcenter *O*) of triangle *ABC*,

$$\angle EBC = \frac{1}{2}\ \overset{\frown}{BC}, \text{ and } \angle DPC\ (=\angle BPC) = \frac{1}{2}\ \overset{\frown}{BC}.$$
Therefore, ∠*EBC* = ∠*DPC*.                                                       (II)

From (I) and (II), by transitivity, ∠*DXC* = ∠*EBC*, which are then alternate-interior angles, making the Simson line *XDZ* parallel to tangent line *EB*.

Another Simson-line curiosity is that if we have two Simson lines generated for the same triangle by two distinct points of the circumscribed circle, the angle formed by the two Simson lines is one-half the measure of the (larger) arc they intercept on the circle.

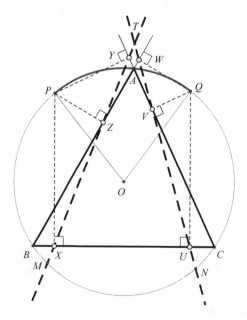

FIGURE 6-5

In figure 6-5, the two Simson lines *YZX* and *UVW* generated by the points *P* and *Q*, respectively, intersect the circle-determining arc $\overset{\frown}{PQ}$. The angle, $\angle MTN$, formed by the two Simson lines has one-half the measure of the intercepted arc $\overset{\frown}{PQ}$.

We also have another interesting property of the Simson line. A Simson line can be shown to bisect the line that joins the orthocenter with the generator point of the Simson line. We can see this in figure 6-6, where point *P* is used to generate the Simson line *XZY* of triangle *ABC*. The line *PH*, joining the orthocenter, *H*, of the triangle with point *P*, is bisected by the Simson line at point *M*, or *PM = HM*.

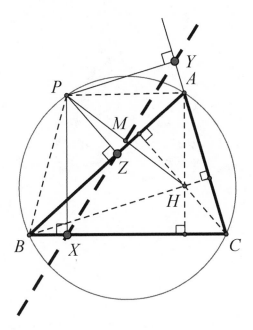

FIGURE 6-6

Using dynamic geometry software—such as the Geometer's Sketchpad® or GeoGebra®—the Simson line can be shown to generate a famous geometric shape called the *Steiner deltoid* or *Steiner's hypocycloid*, named after its founder, Jakob Steiner (1796–1863), who presented it in 1856.[1] When the point *P* moves around the circumscribed circle of triangle *ABC*, the Simson line generates the Steiner deltoid. (See figure 6-7.) The Simson lines are tangent to the newly formed deltoid.

The Steiner deltoid is tangent at three points to the nine-point circle of this triangle, which we will present later in this chapter. Its circumscribed circle is the *Steiner circle*, and its inscribed circle is the nine-point circle. The *Steiner circle* is shown in figure 6-7. The center point of the Steiner circle is $N$, which is also the center point of the nine-point circle, its radius is $\frac{3}{2}R$.

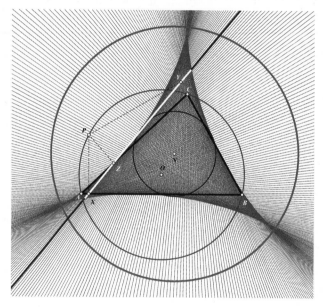

FIGURE 6-7

The Simson line has many other gems that we shall leave to the reader to discover.[2]

## THE EULER LINE

We have already seen a fine collection of significant points of a triangle, such as the orthocenter, the incenter, the centroid, the circumcenter, the

Gergonne point, the Nagel point, the middle point, the Fermat point, the Brocard points, the Miquel point, and symmedian point. However, in high school we already encountered some of these points, such as the orthocenter $H$, the centroid $G$, and the circumcenter $O$. These points enable us to reveal a surprising result: They are collinear! In 1763,[3] the famous Swiss mathematician Leonhard Euler (1707–1783) discovered that these points were truly collinear—as long as the triangle is not equilateral, then they would coincide. (See figure 6-8.) This line of collinear points is therefore called the *Euler line*, $e$.

Dynamic geometry software quickly shows us that the three points are collinear. But this is merely a demonstration. To establish this relationship's truth, we would have to prove the collinearity. This was then Euler's task as well, since he used rather primitive tools to discover this collinearity and then had to develop a proof. As we search for a method of proof, we realize that we cannot use Menelaus's theorem, since the points in question do not lie on the sides of the triangle.

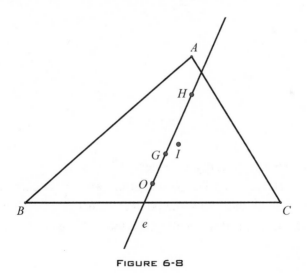

FIGURE 6-8

Furthermore, we can show that the centroid, $G$, of a triangle trisects the segment $HO$, the segment from the orthocenter $H$ to the circumscribed center $O$. (See figure 6-9.)

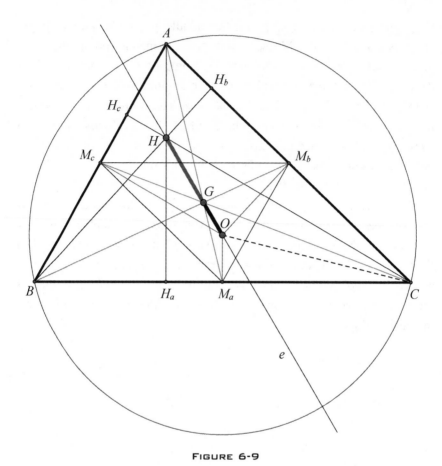

FIGURE 6-9

To prove this collinearity, consider in triangle $ABC$, the line $GO$ (figure 6-10), formed by the centroid ($G$) and the center of the circumscribed circle ($O$), where we will then place the point $P$ so that $\frac{OG}{GP} = \frac{1}{2}$. We will now set out to show that point $P$ is actually the orthocenter, point $H$.

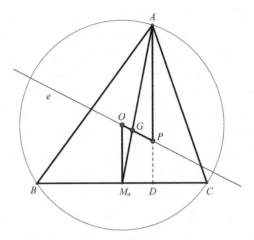

FIGURE 6-10

Recall that in triangle $ABC$ the line segment $AM_a$ is a median and is divided by the centroid so that $\frac{M_aG}{GA} = \frac{1}{2}$. Since $\triangle OGM_a \sim \triangle PGA$ (two pairs of corresponding sides are in proportion and the included angles are equal), we have $OM_a \parallel AP$. Since the line from the center of a circle to the midpoint of a chord is perpendicular to the chord, $OM_a \perp BC$. Therefore, $APD$ is an altitude of the triangle, since $AP \perp BC$. That means that point $P$ is on the altitude $AH_a$ ($D = H_a$). We can repeat this for the other altitudes and then establish that the point $P$ coincides with the orthocenter $H$.[4]

We should also note that the center of the inscribed circle of a triangle ($I$), along with the centroid ($G$), and the Nagel point ($N$) are also collinear.

## THE NINE-POINT CIRCLE

We know that any three noncollinear points lie on a unique circle. Some quadrilaterals have all four of their vertices also on a unique circle. These quadrilaterals we called *cyclic quadrilaterals*. A general parallelogram— other than a square or a rectangle—does not have all of its vertices on a unique circle. For a quadrilateral to have all of its vertices on a circle, the opposite angles must be supplementary, as is the case, for example, with an isosceles trapezoid. However, to find more than four points that lie on the same circle has been a long-standing challenge for mathematicians.

In 1765, Leonhard Euler showed that there are six points of a triangle that lie on a unique circle, namely, the midpoints of the sides, and the feet of the altitudes. Yet not until 1820, when a paper[5] published by the French mathematicians Charles Julien Brianchon (1783–1864) and Jean-Victor Poncelet (1788–1867) appeared, were an additional three points of a triangle added to Euler's circle of six points. These new points were the midpoints of the segments from the orthocenter to the vertices. This paper contained the first complete proof that, in fact, these nine points all lie on the same circle and therefore gave the circle a name — *the nine-point circle*.

The German mathematician Karl Wilhelm Feuerbach (1800–1834) has much of his fame resting on a paper he published in 1822, where he stated that "the circle which passes through the feet of the altitudes of a triangle touches all four of the circles which are tangent to the three sides of the triangle" (*Eigenschaften einiger merkwürdigen Punkte des geradlinigen Dreicks*). Here he was referring to the triangle's inscribed circle and the three escribed circles — those externally tangent to each of the sides of the triangle. (See figures 6-11 and 6-12.) As a result of this work, the theorem is referred to as the *Feuerbach theorem* and the nine-point circle is also sometimes called the *Feuerbach circle*.

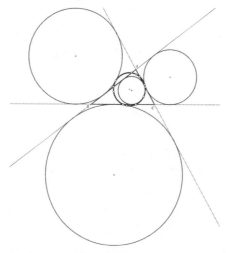

FIGURE 6-11

The thirteen points on the nine-point circle, $c_N$, with the center $N$ are as follows (figure 6-12):

Nine-point Circle, $c_N$ (Euler; Brianchon; Poncelet):
$M_a, M_b, M_c$ – midpoints of the sides of triangle $ABC$,
$H_a, H_b, H_c$ – feet of the altitudes of triangle $ABC$,
$E_a, E_b, E_c$ – the so-called Euler points of triangle $ABC$ (these are the
    midpoints of the segments between the orthocenter $H$ and the ver-
    tices of triangle $ABC$) (Feuerbach):
    $F_i$ – point of tangency with the inscribed circle (at $I$),
    $F_a$ – point of tangency with the escribed circle (at side $a$),
    $F_b$ – point of tangency with the escribed circle (at side $b$),
    $F_c$ – point of tangency with the escribed circle (at side $c$).

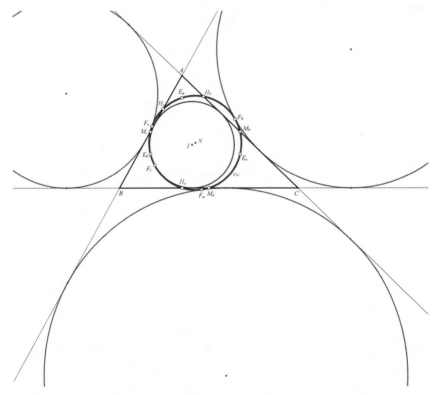

FIGURE 6-12

Roger A. Johnson (1890–1954) in his landmark book writes Feuerbach's famous theorem "is perhaps the most famous of all theorems of the triangle, aside from those known in ancient times."[6] A similar sentiment is shared by the American mathematician Howard Eves (1911–2004).[7]

Let us now begin to justify the placement of these nine points on one circle. We will begin with the first six points. Remember each of the sets of three points will automatically lie on a circle (since they are not collinear). We shall begin by showing that the three midpoints of the sides of the triangle and the foot of one of the altitudes will all lie on the same circle.

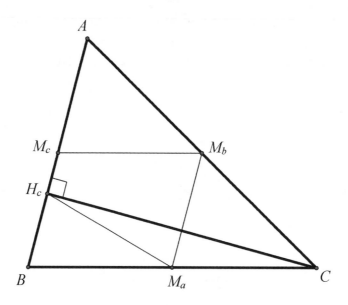

FIGURE 6-13

We begin by trying to establish four points on a circle. In figure 6-13, points $M_a$, $M_b$, and $M_c$ are the midpoints of the three sides, $BC$, $AC$, and $AB$, respectively, of triangle $ABC$. We also have $CH_c$ as an altitude of triangle $ABC$. Since $M_aM_b$ is a midline of triangle $ABC$, $M_aM_b \parallel AB$. Therefore, quadrilateral $M_aM_bM_cH_c$ is a trapezoid. Also, $M_bM_c$ is a midline of triangle $ABC$, so that $M_bM_c = \frac{1}{2}BC$. Since $M_aH_c$ is the median to the hypotenuse of right triangle $BCH_c$, we have $M_aH_c = \frac{1}{2}BC$. Therefore $M_bM_c = M_aH_c$ and trapezoid $M_aM_bM_cH_c$ is isosceles. You will recall that when the opposite angles of a quadrilateral are supplementary, as in the case of an isosceles trapezoid, the quadrilateral is cyclic. Therefore, quadrilateral $M_aM_bM_cH_c$ is cyclic, and we now have established that these four points lie on the same circle.

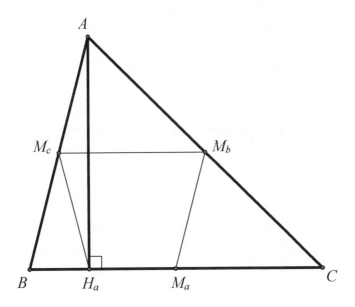

FIGURE 6-14

To simplify matters, we redraw triangle $ABC$ (figure 6-14), this time with altitude $AH_a$. Using the same argument as before, we find that quadrilateral $M_aM_bM_cH_a$ is an isosceles trapezoid and is therefore cyclic. So we now have five points on one circle (i.e., points $M_a$, $M_b$, $M_c$, $H_c$, and $H_a$). By repeating the same argument for altitude $BH_b$, we can then state that points $H_a$, $H_b$, and $H_c$ lie on the same circle as points $M_a$, $M_b$, and $M_c$.

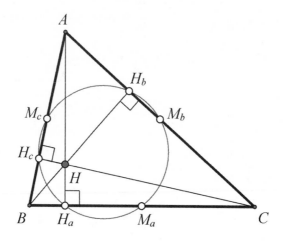

FIGURE 6-15

In figure 6-15 we see the six points ($M_a$, $M_b$, $M_c$, $H_a$, $H_b$, and $H_c$) that Euler established as lying on one circle.

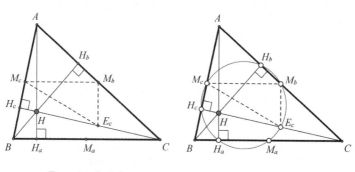

FIGURE 6-16a          FIGURE 6-16b

We will now embark on our quest to establish another three points on the circle that already has six established points on it. In figures 6-16a and 6-16b, we have $H$ as the orthocenter (the point of intersection of the altitudes), and $E_c$ is the midpoint of $CH$. ($E_a$, $E_b$, and $E_c$ are the midpoints of the segments $AH$, $BH$, and $CH$—the so-called Euler points of triangle $ABC$.)

It is this point, $E_c$, that we want to establish on the circle of six points. The line $M_bE_c$, which is a midline of triangle $ACH$, is parallel to $AH$, or essentially to altitude $AH_a$. Since $M_bM_c$ is a midline of triangle $ABC$, it follows that $M_bM_c$ is parallel to $BC$. Therefore, since triangle $AH_aC$ is a right triangle, and the sides of triangle $E_cM_bM_c$ are parallel to the sides of triangle $AH_aC$, then $\angle E_cM_bM_c$ is also a right angle. Thus, quadrilateral $E_cM_bM_cH_c$ is cyclic (opposite angles are supplementary). This places point $E_c$ on the circle determined by points $M_b$, $M_c$, and $H_c$. We now have seven points on the one circle.

We can repeat this procedure with point $E_b$, the midpoint of $BH$ (see figures 6-17a and 6-17b). As before, $\angle M_bM_aE_b$ is a right angle, as is $\angle M_bH_bE_b$. Therefore points $M_b$, $H_b$, $E_b$, and $M_a$ are concyclic (opposite angles are supplementary). We now have $E_b$ as an additional point on our circle, making it an eight-point circle.

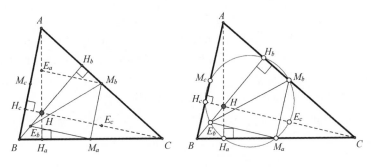

FIGURE 6-17a          FIGURE 6-17b

To locate our final point on the circle, consider point $E_a$, the midpoint of $AH$. As we did earlier, we find $\angle M_aM_bE_a$ to be a right angle, as is $\angle M_aH_aE_a$. Therefore, quadrilateral $M_aH_aE_aM_b$ is cyclic and point $E_a$ is on the same circle as points $M_b$, $M_a$, and $H_a$. We have shown, therefore, that *nine* specific points lie on this circle. (See figure 6-18.)

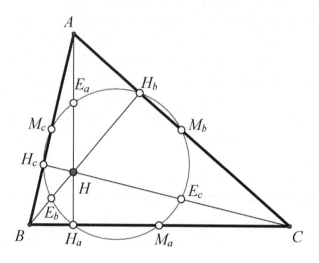

FIGURE 6-18

We have established that there are nine points on the same circle, called the *nine-point circle*, namely, the midpoints of the sides, the feet of the altitudes, and the midpoints of the segments from the orthocenter to the vertices. With a bit of further exploration, we can find some rather novel properties of this circle, some of which we shall identify now.

## SOME PROPERTIES RELATED TO THE NINE-POINT CIRCLE

As we refer to figure 6-19, we will show some rather unexpected properties that this already-unique nine-point circle provides:

- The center, $N$, of the nine-point circle of a triangle is the midpoint of the segment, $HO$, that is, the segment from the orthocenter to the center of the circumscribed circle. This means that the center, $N$, of the nine-point circle lies on the *Euler line*. (See appendix for proof.)
- The centroid of a triangle, $G$, trisects the Euler line segment, $HO$,

the segment from the orthocenter to the center of the circumscribed circle: $HG = 2\ GO$, or $\frac{HN}{NG} = \frac{HO}{GO} = \frac{3}{1}$ . (See p. 144, figure 6-9.)

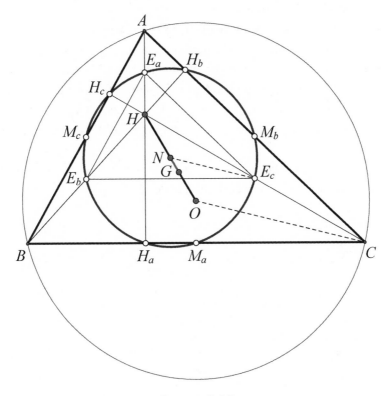

FIGURE 6-19

- The nine-point circle of the triangle $ABC$ is also the circumscribed circle for each of the following triangles:
  - Triangle $M_a M_b M_c$, the triangle formed by the midpoints of the sides of the original triangle.
  - Triangle $H_a H_b H_c$, the triangle formed by the feet of the altitudes.
  - The *Euler triangle* $E_a E_b E_c$.
- The length of the radius of the nine-point circle of a triangle is one-half the length of the radius of the circumscribed circle. This can be seen in figure 6-19, where $NE_c = \frac{1}{2}\ OC$, and in this case also parallel

to it, which helps justify the statement. Six of the points on the nine-point circle are the vertices of its diameter:

$M_aE_a = M_bE_b = M_cE_c (= OC)$. (To justify this, note that $\angle M_aM_bE_a$ and $\angle M_aH_aE_a$ are right angles.)

- All triangles inscribed in a given circle and having a common ortho-center also have the same nine-point circle.
- Tangents to the nine-point circle of a triangle at the midpoints of the sides of the triangle are parallel to the sides of the orthic triangle.

(When we connect the feet of the altitudes, we form a special pedal triangle of the original triangle, a so-called *orthic triangle* $H_aH_bH_c$.)

One such tangent (at point $M_c$) is shown in figure 6-20 to be parallel to side $H_aH_b$ of the orthic triangle $H_aH_bH_c$.

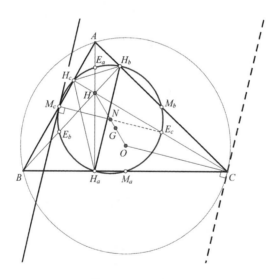

FIGURE 6-20

- Tangents to the nine-point circle at the midpoints of the sides of the given triangle are parallel to the tangents to the circumscribed circle at the opposite vertices of the given triangle. In figure 6-20, the tangent at point $C$ is parallel to the tangent at point $M_c$.
- The four triangles formed by the four points—the vertices

of the triangle and the orthocenter—create an *orthocentric system*, each of whose triangles has the same orthic triangle and the same nine-point circle. (An *orthocentric system* is a set of four coplanar points, each of which is the orthocenter of the triangle formed by the remaining three points.)

* In figure 6-21, the four triangles forming an orthocentric system are:
  – $\triangle ABC$ with the orthocenter $H$,
  – $\triangle AHC$ with the orthocenter $B$,
  – $\triangle BHC$ with the orthocenter $A$, and
  – $\triangle AHB$ with the orthocenter $C$.

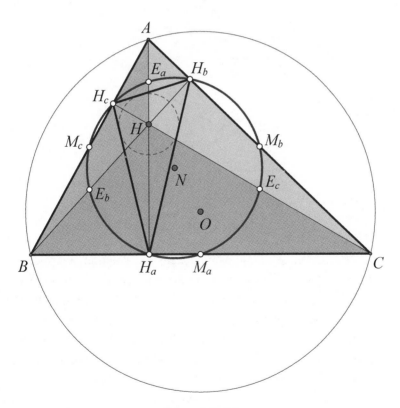

FIGURE 6-21

It is also interesting to note that the radii of the four circumscribed circles of these triangles are equal ($OA = PB = QC = RA$). Furthermore, if we take the centers of these circumscribed circles, we will find that they have the same nine-point circle as the original orthocentric system. These circles are shown in figure 6-22.

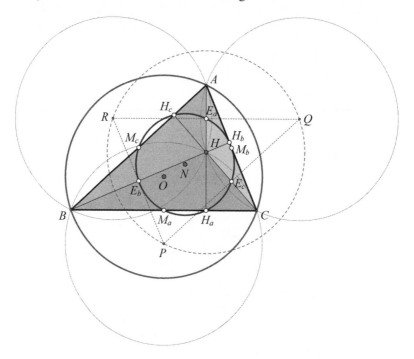

FIGURE 6-22

We can make an interesting extension: The four centroids of the four triangles of an orthocentric system also form an orthocentric system whose nine-point circle is concentric with that of the nine-point circle of the original orthocentric system. We leave this relationship and others to be found here for the reader to discover.

- The nine-point circle $c_N$ of a triangle $ABC$ is tangent to the inscribed circle and the escribed circles of the triangle. This is shown in figure 6-23. (See also figures 6-11 and 6-12, where you will find the names of the thirteen points).

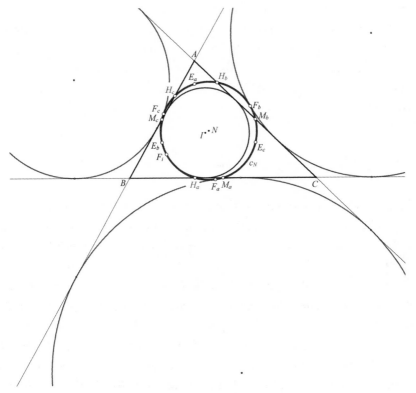

FIGURE 6-23

The last of these properties is — as we indicated earlier — one of the most famous properties of the nine-point circle and was first discovered (and proved) by Karl Wilhelm Feuerbach, a German mathematician, in 1822, and consequently bears his name. This property establishes a relationship between the nine-point circle and the inscribed circle and escribed circles of a triangle.

The point $F_i$ of tangency of the nine-point circle $c_N$ with the inscribed circle is the so-called *Feuerbach point*. Also, the triangle $F_a F_b F_c$ is named after Feuerbach.

The justifications (or proofs) for some of these properties of the nine-point circle can be found in the appendix.

With this all-encompassing circle, we are about to bring our chapter on the various lines, points, and circles of a triangle to a close. We have tried to provide a view of the most dramatic and often-overlooked relationships among the primary components of plane geometry. There are—needless to say—many more to be found. This pleasure we leave to the reader. Yet one more relationship merits mention and is a fine place to leave the reader in wonderment.

## MORLEY'S THEOREM

We will close this chapter on triangle parts with a very famous theorem that was first published in 1900 by Frank Morley (1860–1937) and is one of the more difficult theorems in geometry to prove. (A proof is offered in the appendix.) Yet its beauty lies in the simplicity of the statement: The intersections of the adjacent angle trisectors of any triangle always meet in three points determining an equilateral triangle. Figure 6-24 shows a number of differently shaped triangles with their trisectors drawn, and in each case, the intersections of the adjacent trisectors determine an equilateral triangle. We invite the reader to try this with a variety of differently shaped triangles to see that it holds true for all triangles. Using dynamic geometry software would provide a very dramatic appreciation for this amazing relationship.

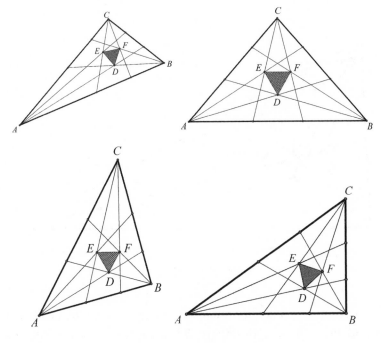

**FIGURE 6-24A, B, C, D**

Beyond Morley's wonderful discovery, you can discover eighteen equilateral triangles under the "Morley's trisector theorem" entry on *Wikipedia* (http://en.wikipedia.org/wiki/Morley%27s_trisector_theorem).

Not to be disappointed by a possible neglect of a concurrency, yes, we have one in this configuration as well. In figure 6-25, we notice that *CD*, *AF*, and *BE* are concurrent.

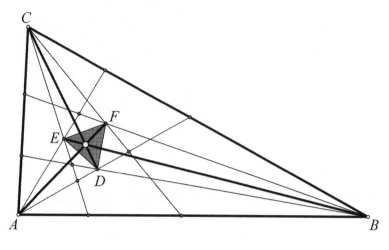

FIGURE 6-25

Concurrency, collinearity, tangency, parallelism, and perpendicularity, along with the many relationships of the various parts of a triangle provide a seemingly endless array of properties that seem to be secretly nested in the triangle. One can continue to endlessly pursue other triangle relationships; however, we feel that we have captured the essence of these delights here in this chapter. After all, we must leave some of these geometric gems for the reader to discover![8]

# CHAPTER 7

# AREAS OF AND WITHIN TRIANGLES

I n the previous chapters, we admired the many fascinating relation-ships that were created by inspecting triangles and the points, lines, and circles that are related to them. In this chapter, we will focus on the many common and surprising *areas* created by triangles and their lines, points, and circles. Triangle areas or their portions hold many fascinating surprises, which we will discover as we journey through this chapter.

Appropriately, we shall begin by first establishing ways in which we can find the area of a triangle. From the most primitive basis, we would begin by considering the area of a square and then the area of a rectangle—that is, if possible by just counting the number of square units they contain, or just multiplying the length by the width. We then progress to the right triangle, which can be seen as half of a rectangle. In figure 7-1, we notice that the right triangle $ABC$ is half the area of rectangle $ABCD$. This allows us to state that the area of a right triangle is one-half the product of its legs. Here, in figure 7-1, $Area\triangle ABC = \frac{1}{2}ab$.

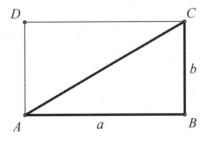

FIGURE 7-1

As we consider the area of a general triangle, we can separate it into two right triangles as shown in figure 7-2. (By "general triangle" we refer to a randomly drawn triangle of no particular shape or property other than being a triangle.) Here we can find the area of each of the two right triangles that compose the given general triangle and then add them to get the area of the complete general triangle, which leads us to the well-known formula for the area of a triangle: one-half the product of a base and the altitude drawn to that base.

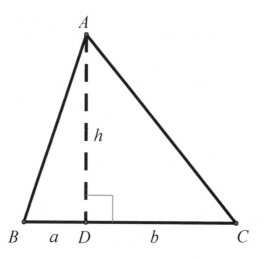

FIGURE 7-2

In figure 7-2, the sum of the areas of the two right triangles ($\triangle ADB + \triangle ADC$) is $\frac{1}{2}ah + \frac{1}{2}bh = \frac{1}{2}h(a+b)$. However, $a + b$ is the base of the general triangle; therefore, the area of triangle $ABC$ is $\frac{1}{2}$(altitude · base).

An analogous development of the formula can be made for an obtuse triangle as shown in figure 7-3.

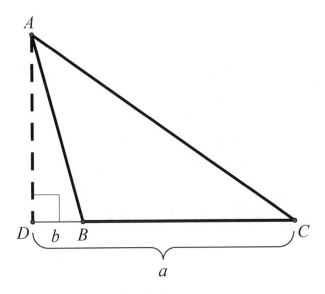

FIGURE 7-3

Here we consider the areas of two overlapping right triangles: triangle *ADC* and triangle *ADB*. The area of triangle *ABC* is equal to the area of triangle *ADC minus* the area of triangle *ADB*. Therefore,

$$Area\ \Delta ABC = \frac{1}{2}ah - \frac{1}{2}bh = \frac{1}{2}h(a-b) = \frac{1}{2}(\text{altitude} \cdot \text{base}).$$

We are not always given such convenient triangle parts to find its area. There may be times when we will be given only the length of two sides of a triangle and the measure of the included angle, as shown in figure 7-4. In this case, we will engage some trigonometry to help us develop a formula to find the area of the triangle. We just established that the area of triangle $ABC = \frac{1}{2}hb$. In right triangle *BCD*, we have $\sin \gamma = \frac{h}{a}$, or $h = a \sin \gamma$. By substituting this value for $h$ into the previously established area formula, we get $Area\ \Delta ABC = \frac{1}{2}(a \text{ in } \gamma) \cdot b = \frac{1}{2}ab \sin \gamma$, which gives us yet another formula for the area of a triangle.

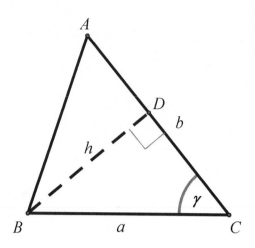

FIGURE 7-4

Were we to have been given the lengths of two sides of a triangle and the measure of an angle *not* included between these two sides, we would not have been able to determine a unique triangle, and hence would be unable to get a definitive area. With such given information, the triangle could be obtuse or acute, thus not allowing us to determine the area of *the* triangle—since it could be either one of two triangles—each with a different area. We show such a case in figure 7-5, where triangle *ABC* and triangle *ABC'* have two pairs of corresponding sides equal, and a pair of corresponding angles (∠*A*) equal—ones not included between these sides in each of the two triangles. Obviously, they have different areas, as one triangle is acute and the other is obtuse.

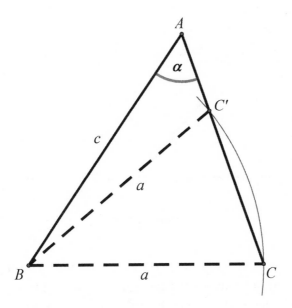

FIGURE 7-5

## COMPARING THE AREAS OF TRIANGLES

The formula *Area* $\triangle ABC = \frac{1}{2}ab$ sin $\gamma$ leads us to a rather unusual way to compare the areas of two triangles that have a pair of corresponding angles equal. Namely, the ratio of the areas of two triangles, having an angle of equal measure, equals the ratio of the products of the two adjacent sides. So that in figure 7-6, where $\angle B = \angle E = \beta$, we have

$$\frac{Area\triangle ABC}{Area\triangle DEF} = \frac{\dfrac{1}{2}ac\sin\beta}{\dfrac{1}{2}df\sin\beta} = \frac{ac}{df}.$$

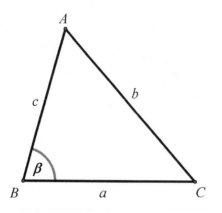
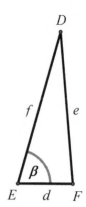

**FIGURE 7-6**

It should be clear that two triangles that have the same lengths of their respective base and altitude have equal areas. Also, two triangles that have the product of their altitude and base the same also have the same area. Furthermore, if two triangles have two sides that have equal products and the angles included between these two sides are equal, then, again, the triangles have equal areas.

Now here is a nice and less well-known relationship: The ratio of the areas of two triangles inscribed in equal circles equals the ratio of the product of their three sides. To justify this bold claim, we first have to consider another triangle relationship, which we show in figure 7-7, where triangle *ABC* is inscribed in a circle with diameter *AD*, and the altitude to side *BC* is *AE*. Then we claim that $AB \cdot AC = AD \cdot AE$.

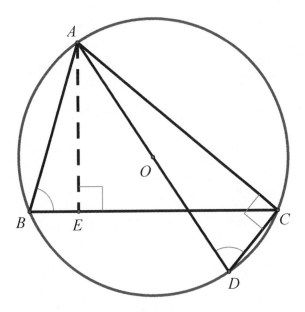

FIGURE 7-7

This can be shown by noting that, since angle $B$ and angle $D$ are both inscribed in the same arc, they are equal. Since triangle $ACD$ is inscribed in a semicircle, it is a right triangle. Thus the right triangles $AEB$ and $ACD$ are similar, and thus $\frac{AB}{AD} = \frac{AE}{AC}$, or $AB \cdot AC = AD \cdot AE$.

We are now ready to justify the comparison of triangle areas mentioned above, namely, that the ratio of the areas of two triangles inscribed in equal circles equals the ratio of the product of their three sides. For figure 7-8, where the two circumscribed circles are equal (each with diameter $d$), we would then want to conclude that

$$\frac{Area\Delta ABC}{Area\Delta PQR} = \frac{a \cdot b \cdot c}{p \cdot q \cdot r} \, .$$

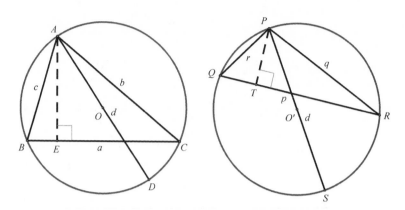

FIGURE 7-8

From the previously demonstrated relationship, we have $b \cdot c = AE \cdot d$, and $q \cdot r = PT \cdot d$.

Therefore, $\frac{b \cdot c}{AE} = d$, and $\frac{q \cdot r}{PT} = d$. Thus, $d = \frac{b \cdot c}{AE} = \frac{q \cdot r}{PT}$, which can be rewritten as $\frac{AE}{PT} = \frac{b \cdot c}{q \cdot r}$.

We are now ready to compare the areas of the two triangles:

$$\frac{Area\Delta ABC}{Area\Delta PQR} = \frac{\frac{1}{2} \cdot a \cdot AE}{\frac{1}{2} \cdot p \cdot PT} = \frac{\frac{1}{2} \cdot a}{\frac{1}{2} \cdot p} \cdot \frac{AE}{PT} = \frac{a}{p} \cdot \frac{AE}{PT} = \frac{a \cdot b \cdot c}{p \cdot q \cdot r}.$$

We know that when we are given the lengths of the three sides of a triangle, a unique triangle is determined. Therefore, in such a case we should be able to establish the area of the triangle. Heron of Alexandria (ca. 10–ca. 70 CE) developed a nifty formula to enable us to find the area of a triangle when the only information that we are given about a triangle is the lengths of its sides. This formula for the area of triangle $ABC$ (figure 7-9) is

$$Area\Delta ABC = \sqrt{s(s-a)(s-b)(s-c)}, \text{ where } s = \frac{a+b+c}{2}, \text{ which is the}$$
semiperimeter of the triangle.

The development of this formula is provided in the appendix. Applying this formula is quite simple. For practice, we shall apply it to a triangle whose sides have lengths 13, 14, and 15 units long, which then has a semiperimeter of 21. Thus, with Heron's formula, we get the area of the triangle as

$$\sqrt{21 \cdot (21-13) \cdot (21-14) \cdot (21-15)} \; = \; \sqrt{21 \cdot 8 \cdot 7 \cdot 6} \; = 84.$$

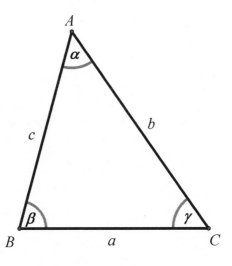

FIGURE 7-9

There are lots of other formulas for finding the area of a triangle, each requiring the measure of various parts of the triangle. We offer just some of these triangle-area formulas here (referring to figure 7-10):

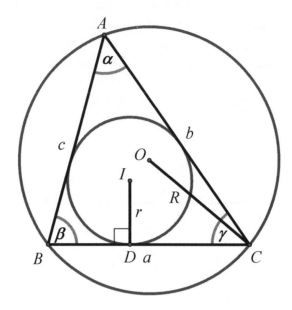

FIGURE 7-10

$Area\triangle ABC = \dfrac{abc}{4R}$   ($R$ = radius of the circumscribed circle)

$Area\triangle ABC = r \cdot s$   ($r$ = radius of the inscribed circle, and $s$ = semiperimeter)

$Area\triangle ABC = \dfrac{\tan\alpha}{4}(b^2 + c^2 - a^2)$   (when $\alpha \neq 90°$)

$Area\triangle ABC = \dfrac{a^2}{2} \cdot \dfrac{\sin\beta \cdot \sin\gamma}{\sin\alpha}$

$Area\triangle ABC = \dfrac{1}{4} \cdot \dfrac{a^2 + b^2 + c^2}{\cot\alpha + \cot\beta + \cot\gamma}$

$Area\triangle ABC = \dfrac{h_a h_b}{2\sin\gamma}$

$Area\triangle ABC = \dfrac{R \cdot h_a h_b}{c}$

$Area\triangle ABC = \dfrac{4}{3}\sqrt{m(m - m_a)(m - m_b)(m - m_c)}$, where $m_a$, $m_b$, and $m_c$ are the medians of $\triangle ABC$, and $m = \dfrac{m_a + m_b + m_c}{2}$.

There are also area formulas for special triangles. We already encountered the right triangle—whose area is simply one-half the product of its legs. For an equilateral triangle, we have two convenient formulas. When we are given the length, $s$, of a side of an equilateral triangle, we have the formula

$$Area\Delta ABC = \frac{\sqrt{3}}{4}s^2 .$$

However, when we are given only the length, $h$, of the altitude of the equilateral triangle, then we can use the formula

$$Area\Delta ABC = \frac{\sqrt{3}}{3}h^2 .$$

## PARTITIONING A TRIANGLE

Now that we have developed ways of finding the area of a triangle, given the measures of its various parts, we can begin to determine how to find the areas of portions of a triangle. For example, when we draw the median of a triangle, we will have divided the triangle into two triangles of equal area. In figure 7-11, where $AD$ is a median of triangle $ABC$, we notice that triangles $ABD$ and $ACD$ have equal bases, $BD$, and $DC$, and share the same altitude $AE$. This clearly implies that they have the same area.

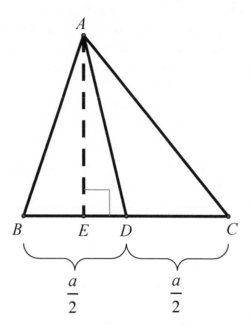

FIGURE 7-11

If in figure 7-12 we draw a cevian from point $A$ to meet $BC$ at a point $D$ one-third the distance from point $B$ to point $C$, then, using the same argument as above, we will have triangle $ABD$ having one-third the area of triangle $ABC$.

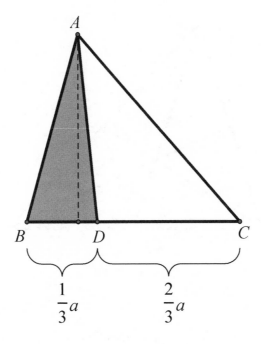

FIGURE 7-12

We are now ready to determine how the three medians of a triangle will partition the triangle's area. Obviously, there will be six triangles formed. But how will the areas of these six triangles compare? Using the case just described above, we can show that the three medians of triangle *ABC* shown in figure 7-13 separate the triangle into six triangles of equal area. To see how this is true, we begin by noting that the area of triangle *ADC* has half the area of triangle *ABC*.

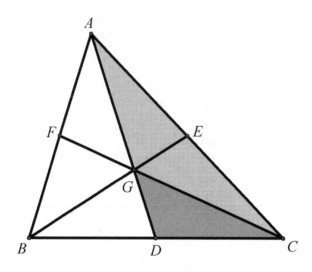

FIGURE 7-13

However, since point $G$ is the trisection point of each of the three medians, we also have

$$Area\Delta DCG = \frac{1}{3}\,Area\Delta ADC = \frac{1}{3}\left(\frac{1}{2}\,Area\Delta ABC\right) = \frac{1}{6}\,Area\Delta ABC.$$

We could repeat this for each of the six triangles shown in the figure. Thus, we can conclude that the medians of a triangle partition the area of the triangle into six triangles of equal area, even though they may have different shapes.

A true surprise now lies before us. If we draw the circumscribed circle for each of the six triangles of equal area we noted in figure 7-13, we will find much to our amazement that their centers all lie on the same circle, as shown in figure 7-14.[1]

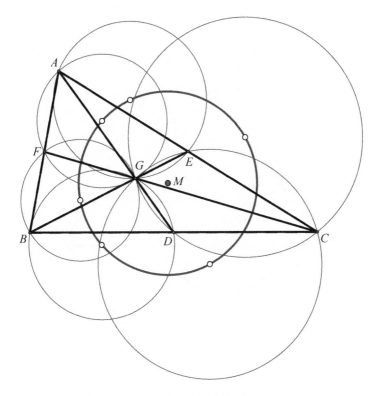

FIGURE 7-14

We now can put these triangle-area findings to use. Suppose we are told the lengths of the three medians of a triangle are 39, 45, and 42 units long. How might we be able to find the area of this triangle?

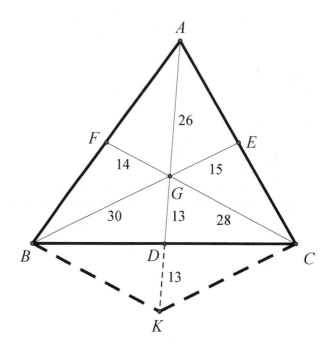

FIGURE 7-15

In figure 7-15, we shall let $AD = 39$, $BE = 45$, and $CF = 42$. Then, because of the trisection property of the centroid, $AG = 26$, $GD = 13$, $BG = 30$, $GE = 15$, $CG = 28$, and $GF = 14$.

We will extend $GD$ its own length to point $K$ (as shown in figure 7-15), so that $GD = DK$. Then quadrilateral $CGBK$ is a parallelogram (since the diagonals bisect each other). We have the side lengths of triangle $CGK$ as 26, 28, and 30 with the semiperimeter 42. Using Heron's formula we get the area of triangle $CGK$ as $\sqrt{42 \cdot (42-26) \cdot (42-28) \cdot (42-30)} = \sqrt{42 \cdot 16 \cdot 14 \cdot 12} = \sqrt{112,896} = 336$.

Yet half the area of this triangle $CGK$ (triangle $GDC$) is 168, which we showed above is one-sixth of the area of the entire triangle. Therefore, the area of triangle $ABC$ is $6 \cdot 168 = 1,008$.

Now that we have a method for partitioning a triangle into six equal-area triangles, we will show how we can partition a triangle into four equal-area triangles. To do that, we simply draw line segments joining the midpoints of the sides of the triangle, as shown in figure 7-16. These four triangles are all congruent to each other (by SSS), since the lines *FE*, *DF*, and *DE* are parallel to the sides of the original triangle; and thus, each is one quarter of the original triangle. The triangle *DEF* is the so-called *medial triangle*.

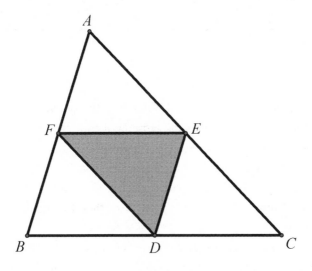

FIGURE 7-16

Using the property that a cevian divides a triangle's area proportional to the segments along the side it partitions, we can now entertain ourselves with lots of geometric configurations that use this property. Take, for example, triangle *ABC* (shown in figure 7-17a), where *D* is the midpoint of *BC*, *E* is the midpoint of *AD*, *F* is the midpoint of *BE*, and *G* is the midpoint of *FC*. We can show that the area of triangle *EFG* is one-eighth of the area of triangle *ABC*. To do this we simply apply our previously established relationship with the median cutting a triangle into two equal areas.

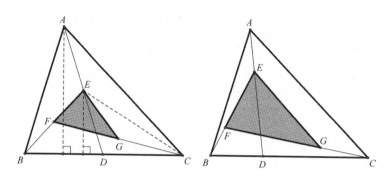

**FIGURE 7-17**

We begin with the altitude of triangle *ABC* drawn to side *BC* and realize that it is twice as long as the altitude of triangle *BEC* drawn to side *BC*. Therefore, since the two triangles share the same base, *Area* Δ*BEC* = $\frac{1}{2}$ *Area* Δ*ABC*. With *CF* as the median of triangle *BEC*, *Area* Δ*EFC* = $\frac{1}{2}$ *Area* Δ*BEC*. Similarly, with *EG* the median of triangle *EFC*, *Area* Δ*EFG* = $\frac{1}{2}$ *Area* Δ*EFC*. Putting this all together, we find that *Area* Δ*EFG* = $\frac{1}{8}$ *Area* Δ*ABC*.

You might want to try other such configurations, such as if *BD* = $\frac{1}{3}$ *BC*, *AE* = $\frac{1}{3}$ *AD*, *BF* = $\frac{1}{3}$ *BE*, and *CG* = $\frac{1}{3}$ *CF* (shown in figure 7-17b), what part of the area of triangle *ABC* would be the area of triangle *EFG*?[2] Try some other variations and see if a pattern emerges.

We partitioned a triangle into two equal areas by drawing a median (*CS*). However, suppose we wish to partition a triangle into two equal areas by drawing a line from a given point (*P*) on a side to a point (*R*, or *R'*) on another side. Using triangle *ABC* in figures 7-18a and 7-18b, we select at random the point *P* on side *AB*, and through that point we would like to draw a line that will cut the triangle into two equal areas (figure 7-18a: for *all* points *P* between *A* and *S*; figure 7-18b: for *all* points *P* between *B* and *S*).

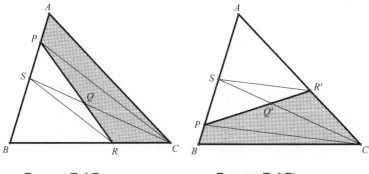

FIGURE 7-18a                    FIGURE 7-18b

Our goal, now, is to show that we can, in fact, draw a line through this randomly selected point $P$ that will separate the triangle $ABC$ into two equal-area regions. First, we will draw the median $CS$. Through point $S$ we draw a line parallel to $CP$, intersecting $BC$ at point $R$ (or intersecting $AC$ at point $R'$ as shown in figure 7-18b). Our claim is that the line $PR$ (or $PR'$ in figure 7-18b) is the line we seek.

Begin with the notion that the median of a triangle separates it into two equal-area triangles. Therefore, we begin our construction (figure 7-18a) by drawing median $CS$, giving us that $Area\ \triangle ACS - \frac{1}{2}\ Area\ \triangle ABC$. Since triangles $PSC$ and $CRP$ have equal altitudes and share the same base ($PC$), they have equal areas. If we subtract the area of triangle $PQC$ from each of these two triangles, we have $Area\triangle PQS = Area\triangle RQC$. Now, $\frac{1}{2}\ Area\ \triangle ABC = Area\ \triangle ACS = Area\triangle PQS + Area\ PQCA = Area\ \triangle RQC + Area\ PQCAS = Area\ APRC$. Therefore, $\frac{1}{2}Area\ \triangle ABC = Area\ APRC = Area\ \triangle BPR$. Thus, $PR$ divides the triangle $ABC$ into two equal areas, which was our original goal. Analogously, for figure 7-18b, we have the same conclusion.

We can also use this technique to *trisect* the area of a triangle by drawing lines through a randomly selected point, $P$, on one of the triangle's sides (see figures 7-19a, 7-19b, and 7-19c). In figure 7-19b, the lines (through point $P$) that trisect the area of triangle $ABC$ are $PR$ and $PS$. To accomplish this, we begin by trisecting $BC$, with points $T$ and $U$. By drawing lines $TR$ and $US$ parallel to $AP$, we will have determined points $R$

and *S* on *AB* and *AC*, respectively. Using a procedure similar to that above, we can show that *PR* and *PS* are the two lines that trisect the area of triangle *ABC*.

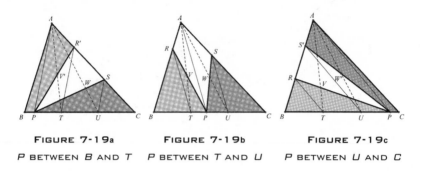

FIGURE 7-19a           FIGURE 7-19b           FIGURE 7-19c

*P* BETWEEN *B* AND *T*    *P* BETWEEN *T* AND *U*    *P* BETWEEN *U* AND *C*

This technique actually enables us to partition a triangle into any number of equal-area regions by drawing lines through one point on the side of the triangle.

Furthermore, we could also use this technique to create a triangle equal in area to a given triangle and sharing a common baseline. More specifically, we would want to construct a triangle equal in area to triangle *ABC*, and having *BD* as its base, where *D* is on *BC*, as shown in figure 7-20.

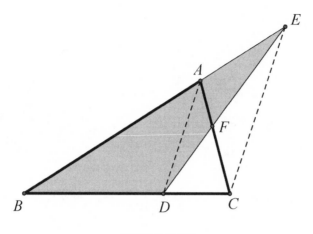

**FIGURE 7-20**

By drawing a line (*CE*) parallel to *AD* and to intersect *BA* (extended) at point *E*, we can form triangle *BDE*, which is equal in area to triangle *ABC*. This is so because *Area* △*AED* = *Area* △*ACD* (same altitude for base *AD*), and we can add triangle *ABD* to each of these two equal-area triangles to get the desired result.

## NAPOLEON'S THEOREM REVISITED

Our earlier study of Napoleon's theorem (see p. 69 and those following) has an extra feature; it has a lovely application to triangle areas. Although it may be a bit complicated by overlapping triangles, the end result is quite astonishing. We consider Napoleon's theorem configuration (with the Fermat point *F*) as shown in figures 7-21a and 7-21b, where the three equilateral triangles are drawn on the sides of the original triangle *ABC*. We will add to this figure parallelogram *AC′CD*. This is done by drawing *AD* parallel, and equal to, *CC′*, which then creates parallelogram *AC′CD*. We then draw *A′D*.

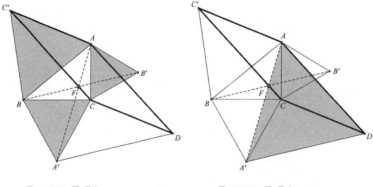

FIGURE 7-21a                    FIGURE 7-21b

We know that triangle $AA'D$ is equilateral, since $AD = AA'$, and $\angle AA'D$ = 60°. (See figure 3-9, p. 74.) Unexpectedly, a rather unusual relationship emerges, namely, that twice the area of triangle $AA'D$ equals the sum of the areas of the three equilateral triangles on the sides of triangle $ABC$, plus three times the area of triangle $ABC$. Symbolically that reads:

$$2 \cdot Area\triangle AA'D = Area\triangle ABC' + Area\triangle BCA' + Area\triangle ACB' + 3 \cdot Area\triangle ABC.$$

Follow along as we justify this rather counterintuitive claim.

$$Area\triangle AA'D = Area\triangle AA'C + Area\triangle ACD + Area\triangle A'CD$$
$$= Area\triangle AA'C + Area\triangle ACC' + Area\triangle AA'B,$$

since $Area\triangle ACD = Area\triangle ACC'$ (parallelogram $AC'CD$), and $Area\triangle A'CD$ = $Area\triangle AA'B$

($\triangle A'CD \cong \triangle AA'B$; because $AB = AC' = CD$, $A'B = A'C$, $AA' = BB' = A'D$).

We also have several triangle congruences in this diagram (figures 7-21a and 7-21b), which, of course, implies equal areas as follows:

$\Delta BB'C \cong \Delta AA'C \Rightarrow Area\Delta BB'C = Area\Delta ACC'$

$\Delta BCC' \cong \Delta AA'B \Rightarrow Area\Delta BCC' = Area\Delta AA'B$

$\Delta ABB' \cong \Delta ACC' \Rightarrow Area\Delta ABB' = Area\Delta ACC'$

This enables us to state the following, which leads to the justification of our rather unexpected original claim.

$2 \cdot Area\Delta AA'D = Area\Delta AA'C + Area\Delta AA'B + Area\Delta ACC' +$
$Area\Delta BB'C + Area\Delta BCC' + Area\Delta ABB'$
$= Area\Delta ABC' + Area\Delta BCA' + Area\Delta ACB' +$
$3 \cdot Area\Delta ABC.$

There are even further area relationships in this configuration to cherish. Consider, once again, Napoleon's theorem. Now we will focus on the equilateral triangle formed by the centers of the three equilateral triangles that were drawn on the sides of the original triangle, which is shown in figure 7-22 (see also figure 3-5, p. 70). This time we have $2 \cdot Area\Delta PQR = Area\Delta ABC + \frac{1}{3}(Area \Delta ABC' + Area\Delta A'BC + Area\Delta AB'C)$.

We leave this justification to the reader.

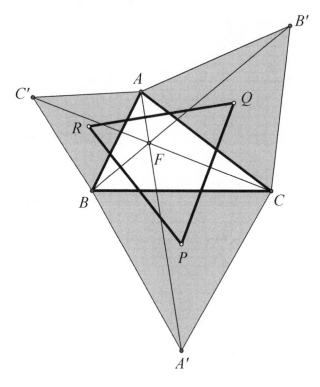

FIGURE 7-22

As we indicated earlier (see figure 3-8, p. 73), there is also an *internal* equilateral triangle—called the *internal Napoleon triangle*—created when the three side-equilateral triangles are placed *internally* on the three sides of the given triangle, that is, overlapping the original triangle *ABC*, as shown in figure 7-23, where triangle *UVW* is the *internal* equilateral triangle, and triangle *PQR* is the *external* equilateral triangle—called the *external Napoleon triangle*. Now here is a true secret of triangles: Who would imagine that the relationship of these two Napoleon triangles is that the difference between the areas of the external and the internal equilateral (Napoleon) triangles is the area of the original triangle? Symbolically we have

$$Area\triangle ABC = Area\triangle PQR - Area\triangle UVW.$$

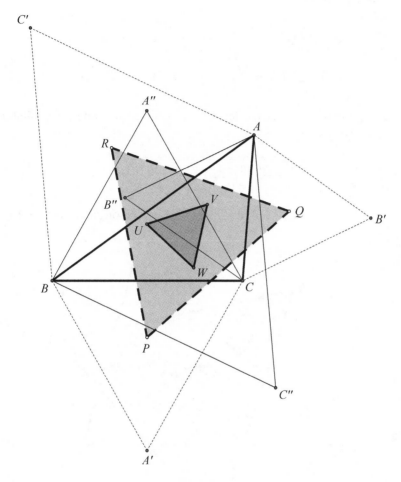

Figure 7-23

To get a clearer picture of this amazing relationship we provide an "unencumbered" picture of this situation in figure 7-24 and some of the related lengths. First, we have in this diagram that the two Napoleon triangles, the external, $\Delta PQR$, and the internal, $\Delta UVW$, have their areas in the relationship

$$Area\Delta ABC = Area\Delta PQR - Area\Delta UVW.$$

These areas are

$$Area\Delta PQR = \frac{Area\Delta ABC}{2} + \frac{\sqrt{3}}{24}(a^2 + b^2 + c^2), \text{ and}$$

$$Area\Delta UVW = \frac{\sqrt{3}}{24}(a^2 + b^2 + c^2) - \frac{Area\Delta ABC}{2}.$$

As a bonus, we offer the length $p$ of each of the sides of equilateral triangle $PQR$ as

$$p = \sqrt{\frac{a^2+b^2+c^2}{6} + \frac{\sqrt{(a+b+c)(a+b-c)(a-b+c)(-a+b+c)}}{2\sqrt{3}}},$$

where $a$, $b$, and $c$ are the lengths of the sides of the original triangle $ABC$.[3]

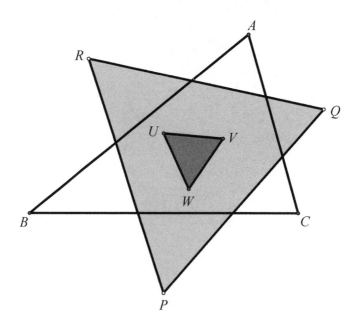

FIGURE 7-24

## INSCRIBED TRIANGLES

We say one triangle is inscribed in another triangle when the vertices of the first triangle are each on the sides of the second triangle. This can be seen in figure 7-25, where we say that triangle *PQR* is inscribed in triangle *ABC*. This leads us to a rather unexpected property: When two triangles are inscribed in a given triangle and are placed so that their vertices are equidistant from the midpoints of the sides of the given triangle, then they have equal areas. This is shown in figure 7-25, where triangles *PQR* and *UVW* are placed with vertices on each of the three sides of triangle *ABC*, with side midpoints so that

$$PM_a = UM_a = x, QM_b = VM_b = y, \text{ and } RM_c = WM_c = z.$$

It then follows that $Area\triangle PQR = Area\triangle UVW$.

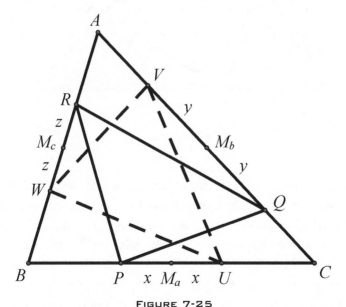

FIGURE 7-25

In figure 7-16, we showed an inscribed triangle situated with vertices on the midpoints of the sides of the original triangle. There the inscribed triangle was earlier shown to be $\frac{1}{4}$ of the area of the original triangle. We can also consider an inscribed triangle that is placed in such a way that the vertices are on the trisection points of the sides of the original triangle as shown in figure 7-26. Here we can show that the area of the inscribed triangle is $\frac{1}{3}$ of the area of the original triangle.

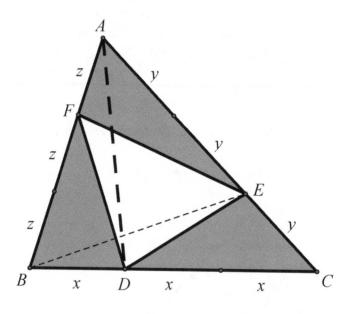

FIGURE 7-26

We can justify this claim in a rather simple way. Recall that a cevian divides the area of a triangle proportional to the segments it determines along the triangle side to which it is drawn. Therefore, the area of triangle *ABD* is one-third the area of triangle *ABC*. However, using similar reasoning, $Area\triangle BDF = \frac{2}{3}Area\triangle ABD$.

Consequently, $Area\triangle BDF = \frac{2}{3} \cdot \frac{1}{3}Area\triangle ABC = \frac{2}{9}Area\triangle ABC$.

A similar argument can be made to show that $Area\triangle CDE = \frac{2}{9}Area\triangle ABC$.

Again, we can show that $Area\triangle AEF = \frac{2}{9}Area\triangle ABC$.

Thus,

$Area\triangle DEF = Area\triangle ABC - (Area\triangle BDF + Area\triangle CDE + Area\triangle AEF)$,

which, with appropriate substitutions, can be restated as $Area\triangle DEF$ $= Area\triangle ABC - (\frac{2}{9}Area\triangle ABC + \frac{2}{9}Area\triangle ABC + \frac{2}{9}Area\triangle ABC) = Area\triangle ABC \cdot$ $\left[1 - \left(\frac{2}{9} + \frac{2}{9} + \frac{2}{9}\right)\right] = \frac{1}{3}Area\triangle ABC$.

Another way to establish that the area of triangle $DEF$ is one-third the area of triangle $ABC$ is to do the following (use cevian $BE$):

$Area\triangle DEF = Area\triangle ADF + Area\triangle ADE - Area\triangle AEF$ $= Area\triangle ABC \cdot \left(\frac{1}{3} \cdot \frac{1}{3} + \frac{2}{3} \cdot \frac{2}{3} - \frac{2}{3} \cdot \frac{1}{3}\right) = \frac{1}{3}Area\triangle ABC$.

The reader might try to show how the area of an inscribed triangle relates to the original triangle, if the vertices of the inscribed triangle are each one-fourth of the distance from each of the original triangle's vertices (consecutively).

## AREAS DETERMINED BY INTERSECTING CEVIANS

We can find some surprising triangle area relationships from the trisection points on the sides of a triangle. In figure 7-27a, we have triangle $ABC$ with trisection points marked along each of its sides. The three cevians $AD, BE$, and $CF$ drawn to the trisection points on the sides of triangle $ABC$, determine triangle $PQR$. It turns out that[4] $Area\triangle PQR = \frac{1}{7}Area\triangle ABC$.

If we now look to figure 7-27b, we can get another triangle ($UVW$) using the cevians $AG, BH$, and $CJ$ that generates the same result. Namely, $Area\triangle UVW = \frac{1}{7}Area\triangle ABC$. In other words, surprisingly, the triangles $PQR$ and $UVW$ have the same area. Moreover, we should note that these two general triangles typically are not congruent, are not similar, and do not have the same perimeter. This adds to the curious relationship of their equal areas.

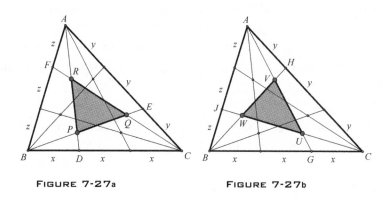

FIGURE 7-27a          FIGURE 7-27b

We can even take this configuration a bit further. Using the trisection points shown in figure 7-28a, it can be shown[5] that the

$$Area\Delta PQR = \frac{1}{25}Area\Delta ABC.$$

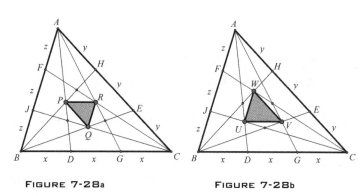

FIGURE 7-28a          FIGURE 7-28b

Joining the other intersection points of the intersecting cevians as shown in figure 7-28b produces triangle *UVW*, which is one-sixteenth of the area of triangle *ABC*, which is stated symbolically as

$$Area\Delta UVW = \frac{1}{16} Area\Delta ABC.$$

Now, unlike the earlier situation where we had two triangles of equal area and their shapes were not at all related, her we have—surprisingly—two triangles *PQR* and *UVW* (shown in figures 7-28a and 7-28b) that are each similar to the original triangle *ABC*, and then, of course, similar to each other—as opposed to the two triangles *PQR* and *UVW* in figures 7-27a and 7-28b, where this was not the case. This can be seen by noticing that the corresponding sides are parallel (e.g. *AB*‖*QR*‖*UW*). We also note that $AB = 5 \cdot QR$, and $AB = 4 \cdot UW$. With this information, we can conclude that $Area \triangle UVW = \frac{25}{16} Area \triangle PQR$.

When we consider the hexagon formed by this side-trisecting cevians (figure 7-29a) we find that the $Area PUQVRW = \frac{1}{10} Area \triangle ABC$. We should note that there is no constant relationship between the perimeters of these two figures.

As if this isn't enough of a surprise, a further relationship exists if we consider other points of intersection of the cevians as shown in figure 7-29b. In this case, $Area KXLYMZ = \frac{13}{49} Area \triangle ABC$. Yet, now we do have a relationship between perimeters. That is, $Perimeter KXLYMZ = \frac{3}{7} Perimeter \triangle ABC$. It is curious that for the hexagon *PUQVRW* we had no relationship of the perimeter to the original triangle, whereas for the hexagon *KXLYMZ* we do have a perimeter relationship.

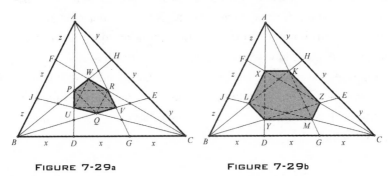

FIGURE 7-29a          FIGURE 7-29b

Now by taking the combination of these two areas, we get a shape shown in figures 7-30a and 7-30b, which are hexagrams—two six-corner stars *P-U-Q-V-R-W* and *K-X-L-Y-M-Z*. (We use the dashes between points to denote hexagrams—as opposed to hexagons.)

It turns out that the $Area\ P\text{-}U\text{-}Q\text{-}V\text{-}R\text{-}W = \frac{7}{100} Area \triangle ABC$ (figure 7-30a), and

$$Area\ K\text{-}X\text{-}L\text{-}Y\text{-}M\text{-}Z = \frac{13}{70} Area \triangle ABC \text{ (figure 7-30b).}$$

Once again satisfying our curiosity about the relationship between perimeters—if one exists—we can conclude that *Perimeter P-U-Q-V-R-W* = $\frac{3}{10}$ *Perimeter*$\triangle ABC$. We cannot make any such claim for the hexagram *K-X-L-Y-M-Z*.

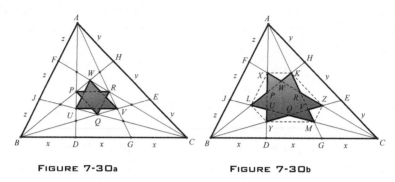

FIGURE 7-30a                    FIGURE 7-30b

So you can see that the side-trisecting cevians provide some rich triangle-area applications, the justifications for these we leave to the ambitious reader. One might also consider how the above relationships manifest themselves when the original triangle is equilateral. Another tack to take in pursuing further investigations is to consider cevians drawn to points that partition each side into four or more equal parts. There are many pursuits one can take for further explorations here.

Suppose you have drawn two medians and a side trisector from the third vertex—as shown in figure 7-31. Here we have, interestingly enough, the triangle formed in the original triangle having $\frac{1}{60}$ of the area of the original triangle.

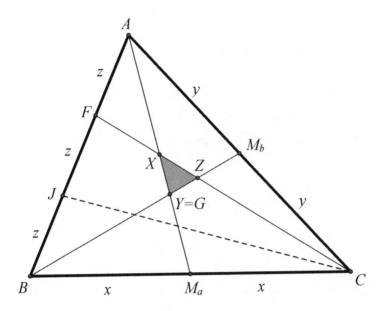

**FIGURE 7-31**

We show this in figure 7-31, where points $M_a$ and $M_b$ are midpoints of their respective sides of triangle $ABC$, and point $F$ is a trisection point of side $AB$. Because $AM_a$ and $BM_b$ are two medians, their intersection point $Y$ is the centroid $G$ of the triangle $ABC$. It can be shown that

$$Area\triangle XYZ = \tfrac{1}{60} Area\triangle ABC.$$

The proof of this unexpected relationship is based on Routh's theorem[6] (1896). This states that for triangle $XYZ$ (figure 7-32), formed by the intersecting cevians in triangle $ABC$,

we have $\frac{AF}{FB} = r$, $\frac{BD}{DC} = s$, and $\frac{CE}{EA} = t$. It follows that

$$\frac{Area\triangle XYZ}{Area\triangle ABC} = \frac{(rst-1)^2}{(rs+r+1)(rt+t+1)(st+s+1)}.$$

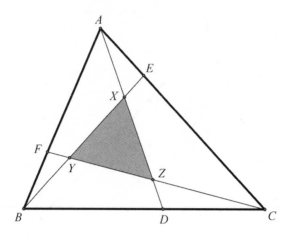

FIGURE 7-32

Applying this to the figure 7-31, we get $\dfrac{AF}{FB} = \dfrac{\frac{1}{3}AB}{\frac{2}{3}AB} = r = \dfrac{1}{2}$,

$\dfrac{BD}{DC} = \dfrac{BM_a}{M_aC} = \dfrac{\frac{1}{2}BC}{\frac{1}{2}BC} = s = 1$, and $\dfrac{CE}{EA} = \dfrac{CM_b}{M_bA} = \dfrac{\frac{1}{2}AC}{\frac{1}{2}AC} = t = 1$.

Therefore, $\dfrac{Area\triangle XYZ}{Area\triangle ABC} = \dfrac{(\frac{1}{2}-1)^2}{(\frac{1}{2}+\frac{1}{2}+1)(\frac{1}{2}+1+1)(1+1+1)} = \dfrac{\frac{1}{4}}{2\cdot\frac{5}{2}\cdot 3} = \dfrac{1}{60}$.

We can also see Ceva's theorem (see chapter 2, p. 43) as a special case of the Routh's theorem. If the cevians $AD$, $BE$, and $CF$ meet at a common point (i.e., $X = Y = Z$), then the area of triangle $XYZ$ is zero, then we can conclude by Routh's theorem that $\dfrac{AF}{FB} \cdot \dfrac{BD}{DC} \cdot \dfrac{CE}{EA} = 1$, which is Ceva's theorem.

It is always nice to see these consistencies throughout mathematics!

Many more such interior triangles can be formed with the various cevian partitions of a triangle.[7] Finding the area relationships, and establishing their truth, can be challenging, but surely worth the effort! We encourage the reader to embark on this rewarding path.

We can conclude this chapter by showing how we can find a triangle equal in area to a given polygon. We will apply this unusual procedure to a pentagon, but surely it can be used with other polygons as well.

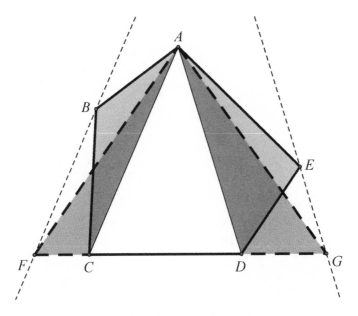

FIGURE 7-33

To construct a triangle equal in area to pentagon *ABCDE* (shown in figure 7-33), we begin by drawing *GE*‖*AD*, and *FB*‖*AC*. Since triangles *AED* and *AGD* share the same base (*AD*), and have equal altitudes to that base, they are equal in area. Similarly, triangles *ABC* and *AFC* are also equal in area. Thus, with appropriate replacements, we find that triangle *AFG* is equal in area to pentagon *ABCDE*.

We have now covered the area of a triangle from its definition and the basic formula through many other formulas that are not too well known. Yet with the above technique that enables us to construct a triangle equal to the area of a given polygon, we can then find the area of an irregular polygon quite easily. This is a fine way to expose a well-kept secret of the triangle's power.

# CHAPTER 8

# TRIANGLE CONSTRUCTIONS

We are now about to embark on a topic in geometry that is perhaps one of the most genuine forms of problem solving in the field. The construction of triangles, given only certain parts of a triangle, can be easy or quite challenging, depending on the information given. In this chapter, we will provide a broad presentation of the beauty of problem solving in geometry through the construction of triangles.

Surely, everyone knows how to draw a triangle. For example, given the lengths of the three sides of a triangle, the triangle can easily be constructed. The question that then arises is what information (minimally) is needed about a triangle in order to be able to construct it. Perhaps, more specifically, how many parts, and which parts, of a triangle must be known in order to be able to construct a triangle. Naturally, we could be given insufficient information—that is, not enough information to determine the triangle. We could also be given more information than is necessary. We then have to decide what is necessary and what may be superfluous.

To determine a unique line we need simply two distinct points. When we have three noncollinear points we can determine a unique triangle—or for that matter a unique circle. We can also determine a unique-size circle if we are given the length of the radius, just as we can determine a unique-size square if we are given the length of a side of the square, or a unique-size equilateral triangle is we are given the length of a side of the triangle. It is interesting to note that to determine an isosceles triangle one needs two bits of information (side length and base, side length and the vertex angle, or base length and vertex angle); for a general triangle, we need three bits of information; and four bits are required to determine a general trapezoid.

Having roughly set the ground rules for what a construction may require, we should next discuss the tools that we can use for these constructions. These tools are an unmarked straightedge and a pair of compasses. (Just as we have a "pair of scissors," so do we also have a *"pair of compasses."* And, by the way, a *compass* is an instrument to determine direction.) All constructions must be exact and no "approximations" are acceptable for these geometic constructions—using these so-called Euclidean tools, with which we can draw only lines connecting two determined points and circles with a given center and a given radius. We cannot measure length as with a ruler, only with compasses, by preserving a radius length.

These constructions are called Euclidean constructions because they stem from the Greek philosopher Plato (428/427–348/347 BCE) and were used extensively in the axioms and postulates at the outset of Euclid's (ca. 360–ca. 280 BCE) monumental work, *Elements*.

## CONSTRUCTIONS OF REGULAR POLYGONS

Before we embark on our journey through the construction of triangles, we ought to make a brief visit to the construction of regular polygons—that is, polygons with equal length sides and equal angles. The most basic of the regular polygons is (naturally) the equilateral triangle, whose construction is done by drawing two equal circles with a common radius as shown in figure 8-1. The next regular polygon is the square that can be done with two equal circles sharing a common radius, then constructing perpendicular lines at the centers, and then completing the square by joining the points at which the perpendiculars meet the circles, as shown in figure 8-2.

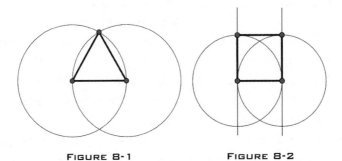

FIGURE 8-1          FIGURE 8-2

The constrcution of a regular pentagon is somewhat more complicated and is tied in with the golden section.[1] In contrast, the construction of the regular hexagon is quite simple. All that one has to do is copy the radius of the (circumscribed circle) hexagon five times around the original circle, as shown in figure 8-3.

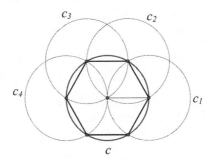

FIGURE 8-3

The next regular polygon that we would seek to construct is one of seven sides. This has been tried—unsuccessfully—for over two thousand years. While pondering in bed on the morning of March 29, 1796, the nineteen-year-old Carl Friedrich Gauss (1777–1855) had an inspiration. He suddenly realized that a polygon of $n$ sides can only be constructed with straightedge and compasses if $n = 2^r \cdot p_1 \cdot p_2 \cdot \ldots \cdot p_m$ (with $r \geq 0$; $m \geq 0$; $r$ and $m$ are natural numbers), where $p_i$ is a prime number of the form $2^{2^k} + 1$ ($i = 1, 2, \ldots, m$; $k \geq 0$; $i$ and $k$ are natural numbers).

Whereupon, we now can state that a regular polygon of seven sides cannot be constructed with straightedge and compasses, since $n = 7$ will not be written as $n = 2^r \cdot p_1 \cdot p_2 \cdot \ldots \cdot p_m$. Seven is not alone here, since the following numbers also do not conform to Gauss's rule: 9, 11, 13, 14, 18, 19, 21, 22, 23, 25, 26, 27, 28, 29, 31, 33, 35, 36, 37, 38, 39, 41, 42, 43, 44, 45, 46, 47, 49, 50, .... Therefore, polygons with such number of sides are not constructible with our Euclidean tools.

If we let $k = 2$ and $r = 0$, we get a seventeen-sided regular polygon, and that is, therefore, constructible, since for $n = 17$: $p_1 = 2^{2^2} + 1 = 16 + 1 = 17$. This discovery, which he came upon through a number of theoretic considerations, had put to rest a problem that has plagued mathematicians for over two thousand years. Gauss did not provide a description of the construction. This he left for others to do. His diary was found in 1899 and begins with this discovery—of which he was most proud. Several years after his monumental discovery, he succumbed to the pressure of a friend and provided a possible method of construction of a seventeen-sided regular polygon.[2] Although Gauss requested that his gravestone have a seventeen-sided regular polygon, a statue of Gauss in Braunschweig, Germany, rests on a seventeen-sided regular polygon.

There is a cute story of a student at the University of Königsberg (today Kaliningrad, Russia) who was pursuing his doctorate and was challenged by his professor to construct a regular polygon of 65,537 sides—using the Euclidean tools (an unmarked straightedge and compasses). He knew the construction was possible because it fit Gauss's formula: Since $65{,}537 = 2^{2^4} + 1 = 2^{16} + 1$ (where $k = 4$) is a prime number. In 1879, some ten years after he was given the challenge, the student arrived with a suitcase containing 250 densely written large sheets of paper that contained the required work. Were one to construct a regular polygon of 65,537 sides in a circle of radius 10 meters, we would find that any two adjacent vertices of the polygon would be about 1 millimeter apart. Yes, the student did receive his doctorate, since nobody had neither the time, nor the inclination, to read through this work. The box containing these pages is still in the mathematics library at the University of Göttingen.[3]

## TRIANGLE CONSTRUCTIONS

In order to construct a triangle with an unmarked straightedge and compasses, we require the measure of three parts of the triangle. For example, if we have the lengths of the three sides $a, b$, and $c$, of the triangle we could construct the triangle (assuming, of course, that the triangle inequality is upheld, namely, that $a + b > c$, $b + c > a$, and $c + a > b$).

The parts of the triangle $ABC$ (shown in figure 8-4) we will initially consider are as follows:

Sides: $a, b, c$

Angles at vertices $A, B$, and $C$, respectively: $\alpha, \beta, \gamma$

Altitudes to sides $a, b$, and $c$, respectively: $h_a, h_b, h_c$

Medians to sides $a, b$, and $c$, respectively: $m_a, m_b, m_c$

Angle bisectors of angles $\alpha, \beta, \gamma$, drawn to sides $a, b$, and $c$, respectively: $t_a, t_b, t_c$

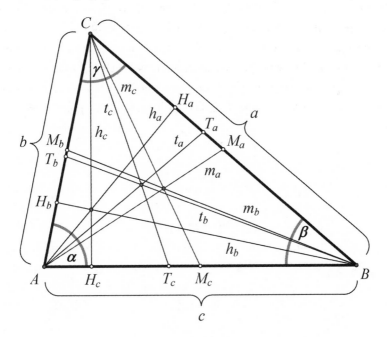

FIGURE 8-4

To further complicate matters, as you noticed with the above mention of the three sides of a triangle, it will not always be possible to construct a triangle with three such given parts. Only a certain combination of these three parts will generate a unique triangle. For example, if we are given the three angles of a triangle, then there are countless triangles that can be constructed—all similar to each other. Furthermore, being given the set $a$, $b$, and $t_a$ will also not allow a triangle to be determined since the parts are interdependent.

If we consider the 15 triangle parts listed above, we find that there are 455 possible combinations.[4] There are, naturally, lots of repetitions included in this number. For example, if we list the set $(a, b, \alpha)$, that is essentially the same as having listed $(b, c, \beta)$, as well as $(c, a, \gamma)$, and the rest of these triples: $(a, b, \beta), (b, c, \gamma)$, and $(c, a, \alpha)$. That is, for one kind of triangle information triple, given two sides and the angle opposite one of them, we have six listed in our 455 possible combinations.

Considering these "repetitions," we actually reduce our number of possible triangle-construction possibilities from 455 to 95. There are 60 triples of six "repetitions." There are 30 triples of three "repetitions," and 5 triples that are singular.[5] Of these 95 possibilities, there are 63 that can be done, 30 that have no possibility for construction, such as the earlier mentioned triple $(a, b, t_a)$; as well as the triple $(h_a, m_b, t_c)$; and two that are underdefined, such as $(a, b, h_c)$, where $h_c = a \cdot \sin \beta$, and $(\alpha, \beta, \gamma)$, where $\alpha + \beta + \gamma = 180°$.

| | Given Data | Number of the Same Type | Constructible | | Given Data | Number of the Same Type | Constructible |
|---|---|---|---|---|---|---|---|
| 1 | $(a, b, c)$ | 1 | yes | 49 | $(\alpha, \beta, h_a)$ | 6 | yes |
| 2 | $(a, b, \alpha)$ | 6 | yes | 50 | $(\alpha, \beta, h_c)$ | 3 | yes |
| 3 | $(a, b, \gamma)$ | 3 | yes | 51 | $(\alpha, \beta, m_a)$ | 6 | yes |
| 4 | $(a, b, h_a)$ | 6 | yes | 52 | $(\alpha, \beta, m_c)$ | 3 | yes |
| 5 | $(a, b, h_c)$ | 3 | yes | 53 | $(\alpha, \beta, t_a)$ | 6 | yes |
| 6 | $(a, b, m_a)$ | 6 | yes | 54 | $(\alpha, \beta, t_c)$ | 3 | yes |
| 7 | $(a, b, m_c)$ | 3 | yes | 55 | $(\alpha, h_a, h_b)$ | 6 | yes |
| 8 | $(a, b, t_a)$ | 6 | no | 56 | $(\alpha, h_a, m_a)$ | 3 | yes |
| 9 | $(a, b, t_c)$ | 3 | yes | 57 | $(\alpha, h_a, m_b)$ | 6 | yes |
| 10 | $(a, \alpha, \beta)$ | 6 | yes | 58 | $(\alpha, h_a, t_a)$ | 3 | yes |
| 11 | $(a, \alpha, h_a)$ | 3 | yes | 59 | $(\alpha, h_a, t_b)$ | 6 | no |
| 12 | $(a, \alpha, h_b)$ | 6 | yes | 60 | $(\alpha, h_b, h_c)$ | 3 | yes |
| 13 | $(a, \alpha, m_a)$ | 3 | yes | 61 | $(\alpha, h_b, m_a)$ | 6 | yes |
| 14 | $(a, \alpha, m_b)$ | 6 | yes | 62 | $(\alpha, h_b, m_b)$ | 6 | yes |
| 15 | $(a, \alpha, t_a)$ | 3 | yes | 63 | $(\alpha, h_b, m_c)$ | 6 | yes |
| 16 | $(a, \alpha, t_b)$ | 6 | no | 64 | $(\alpha, h_b, t_a)$ | 6 | yes |
| 17 | $(a, \beta, \gamma)$ | 3 | yes | 65 | $(\alpha, h_b, t_b)$ | 6 | yes |
| 18 | $(a, \beta, h_a)$ | 6 | yes | 66 | $(\alpha, h_b, t_c)$ | 6 | no |
| 19 | $(a, \beta, h_b)$ | 6 | yes | 67 | $(\alpha, m_a, m_b)$ | 6 | yes |
| 20 | $(a, \beta, h_c)$ | 6 | undetermined | 68 | $(\alpha, m_a, t_a)$ | 3 | yes |
| 21 | $(a, \beta, m_a)$ | 6 | yes | 69 | $(\alpha, m_a, t_b)$ | 6 | no |
| 22 | $(a, \beta, m_b)$ | 6 | yes | 70 | $(\alpha, m_b, m_c)$ | 3 | yes |
| 23 | $(a, \beta, m_c)$ | 6 | yes | 71 | $(\alpha, m_b, t_a)$ | 6 | no |
| 24 | $(a, \beta, t_a)$ | 6 | no | 72 | $(\alpha, m_b, t_b)$ | 6 | no |
| 25 | $(a, \beta, t_b)$ | 6 | yes | 73 | $(\alpha, m_b, t_c)$ | 6 | no |
| 26 | $(a, \beta, t_c)$ | 6 | yes | 74 | $(\alpha, t_a, t_b)$ | 6 | no |
| 27 | $(a, h_a, h_b)$ | 6 | yes | 75 | $(\alpha, t_b, t_c)$ | 3 | no |
| 28 | $(a, h_a, m_a)$ | 3 | yes | 76 | $(h_a, h_b, h_c)$ | 1 | yes |
| 29 | $(a, h_a, m_b)$ | 6 | yes | 77 | $(h_a, h_b, m_a)$ | 6 | yes |
| 30 | $(a, h_a, t_a)$ | 3 | yes | 78 | $(h_a, h_b, m_c)$ | 3 | yes |
| 31 | $(a, h_a, t_b)$ | 6 | no | 79 | $(h_a, h_b, t_a)$ | 6 | no |
| 32 | $(a, h_b, h_c)$ | 3 | yes | 80 | $(h_a, h_b, t_c)$ | 3 | yes |
| 33 | $(a, h_b, m_a)$ | 6 | yes | 81 | $(h_a, m_a, m_b)$ | 6 | yes |
| 34 | $(a, h_b, m_b)$ | 6 | yes | 82 | $(h_a, m_a, t_a)$ | 3 | yes |
| 35 | $(a, h_b, m_c)$ | 6 | yes | 83 | $(h_a, m_a, t_b)$ | 6 | no |
| 36 | $(a, h_b, t_a)$ | 6 | no | 84 | $(h_a, m_b, m_c)$ | 3 | yes |
| 37 | $(a, h_b, t_b)$ | 6 | yes | 85 | $(h_a, m_b, t_a)$ | 6 | yes |
| 38 | $(a, h_b, t_c)$ | 6 | yes | 86 | $(h_a, m_b, t_b)$ | 6 | no |
| 39 | $(a, m_a, m_b)$ | 6 | yes | 87 | $(h_a, m_b, t_c)$ | 6 | no |
| 40 | $(a, m_a, t_a)$ | 3 | yes | 88 | $(h_a, t_a, t_b)$ | 6 | no |
| 41 | $(a, m_a, t_b)$ | 6 | no | 89 | $(h_a, t_b, t_c)$ | 3 | no |
| 42 | $(a, m_b, m_c)$ | 3 | yes | 90 | $(m_a, m_b, m_c)$ | 1 | yes |
| 43 | $(a, m_b, t_a)$ | 6 | no | 91 | $(m_a, m_b, t_a)$ | 6 | no |
| 44 | $(a, m_b, t_b)$ | 6 | no | 92 | $(m_a, m_b, t_c)$ | 3 | no |
| 45 | $(a, m_b, t_c)$ | 6 | no | 93 | $(m_a, t_a, t_b)$ | 6 | no |
| 46 | $(a, t_a, t_b)$ | 6 | no | 94 | $(m_a, t_b, t_c)$ | 3 | no |
| 47 | $(a, t_b, t_c)$ | 3 | no | 95 | $(t_a, t_b, t_c)$ | 1 | no |
| 48 | $(\alpha, \beta, \gamma)$ | 1 | undetermined | | | | |

FIGURE 8-5

Occasionally, we might also be given the radius, $R$, of the circumscribed circle of the triangle; the radius, $r$, of the inscribed circle; or the radii, $r_a, r_b$, and $r_c$, of the three escribed circles of the triangle. This, naturally, increases the number of possible triangle constructions.

Now that we have the parameters of the construction possibilities for creating triangles from three pieces of information, we need to develop an efficient procedure to do the actual constructions. As we mentioned earlier, triangle constructions of this sort are fine examples of genuine problem-solving experiences. An effective technique to approach these construction problems is to begin with a sketch of the constructed triangle and search for parts that can determine fixed portions of the triangle to be constructed. We will do a fair number of these constructions to provide the reader with an opportunity to develop this technique. Remember, there are sometimes alternate ways to do a construction — just as there are often alternate methods to solve many problems. We offer those that we feel are most easily understood. After a construction and its justification, we will also discuss aspects of the construction, such as uniqueness or alternate solutions.

Our first group of triangle constructions (examples 1–17) will involve the previously mentioned triangle parts, namely, sides, angles at vertices, altitudes, medians, and angle bisectors. Later we will also include the radii of related circles (examples 18 and 19).

## EXAMPLE 1: CONSTRUCT TRIANGLE $ABC$, GIVEN ($a$, $b$, $c$)

(We will be using the triple notation "$(x, y, z)$" to denote the three parts of triangle given for a construction problem. In this case, we are given the three side lengths of triangle $ABC$.)

In figure 8-6, we show the problem. We are given the three side lengths and are required to construct the triangle.

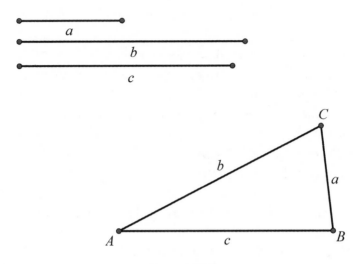

FIGURE 8-6

We describe the construction in figure 8-7, where we show the steps in the encircled numbers. Circles are described by their center and radius length. For a circle with center $A$, and radius $b$, we write this as $c(A,b)$.

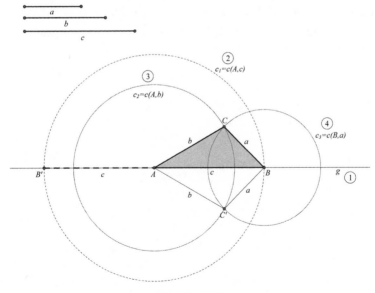

FIGURE 8-7

The construction steps referred to in figure 8-7 are as follows:

① Draw line $g$ and point $A$ on line $g$.

② Construct circle $c_1 = c(A,c)$ with center at $A$ and radius $c$, intersecting line $g$ at $B$.

(In short form, we can write this as $\{B\} = g \cap c_1$ [The symbol $\cap$ means intersection].)

③ Construct circle $c_2 = c(A,b)$.

④ Construct circle $c_3 = c\ (B,a)$.

The intersection, $C$, of the circles $c_2$ and $c_3$ will give us the required triangle: $\triangle ABC$.

(In short form, we can write this as $\{C\} = c_2 \cap c_3$.)

A second point of intersection of the circles is point $C'$, which determines a congruent triangle below the line $g$.

We should note that here, as in most other constructions, the size of the given data is essential in determining if, in fact, such a requested triangle actually exists. This can be seen easily in the present example. Figure 8-8 shows what would happen if the inequality $a + b > c$ does not hold. This is true for any three sides: The sum of any two must be greater than the third, or else the triangle will not exist. That is, $a + b > c$, $b + c > a$, and $c + a > b$.

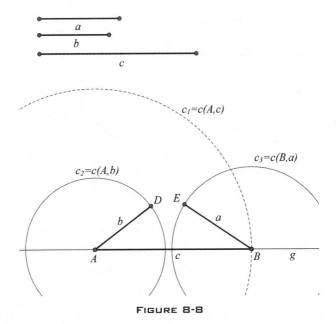

FIGURE 8-8

## EXAMPLE 2: CONSTRUCT A TRIANGLE $ABC$, GIVEN ($a$, $b$, $\angle A$)

This construction calls for a triangle $ABC$, given two sides, $a$ and $b$, and the angle, $\alpha$, opposite one of the given sides. This is shown in figure 8-9; yet side $BC$ could also be on the other side of altitude $CH_c$, as we will see when we do the construction in figure 8-10.

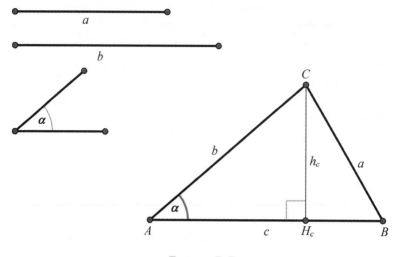

FIGURE 8-9

Remember, as we proceed through the construction, we strive to take fixed or determined measures and craft them together toward our intended construction. Here we have the length $b$ and the angle $\alpha$.

We then need to see what sets of points the length $a$ can assume—that is, in this case, it is the set of points, or locus, that is the circle with center at $C$ and radius $a$.

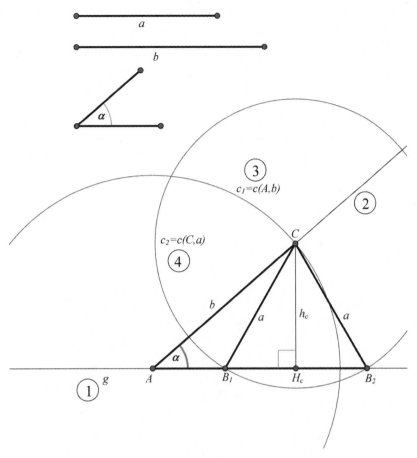

FIGURE 8-10

The step-by-step construction is shown in figure 8-10.

① Begin by drawing line $g$ and placing point $A$ on line $g$.

② Construct the angle $\alpha$ on line $g$ at the point $A$.

③ Along the other ray of angle $\alpha$ mark off length $b$ (by drawing circle $c_1 = c(A, b)$ to create $AC$).

④ By drawing circle $c_2 = c(C, a)$, we find there are two points of intersection with $g$, giving us points $B_1$ and $B_2$.

Therefore, the given information determined two triangles: $\triangle AB_1C$ and $\triangle AB_2C$. If $CH_c$, or $h_c$, is longer than $a$, then there would be no possible triangle fomed, as you can see in figure 8-11. If $CH_c$ is shorter than $a$, then two possible triangles can be formed, as you can see in figure 8-10.

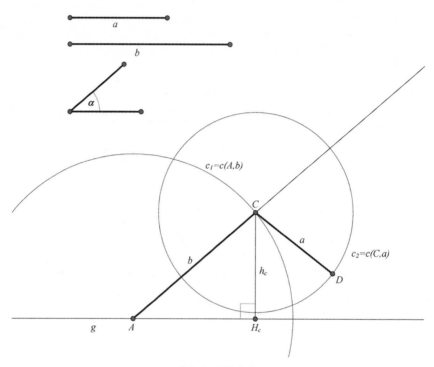

FIGURE 8-11

Should $h_c$ be equal to $a$, then only one triangle can be constructed—a right triangle, as shown in figure 8-12.

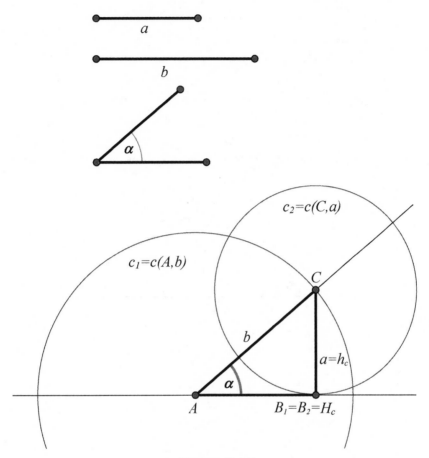

FIGURE 8-12

On the other hand, if $a$ is greater than both $b$ and $h_c$, then we will find two noncongruent triangles ($\triangle AB_1C$ and $\triangle AB_2C$) are constructed, as shown in figure 8-13. But $\triangle AB_1C$ is not a solution, since $\angle B_1AC \neq \alpha$.

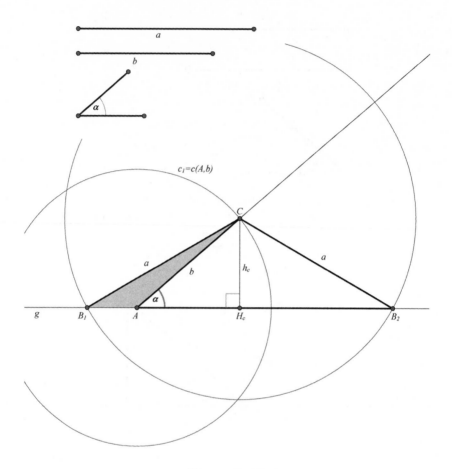

FIGURE 8-13

## EXAMPLE 3: CONSTRUCT A TRIANGLE $ABC$, GIVEN ($a$, $b$, $h_a$)

This construction calls for determining the triangle when two sides, $a$ and $b$, are given as well as the altitude $h_a$ to one of these sides.

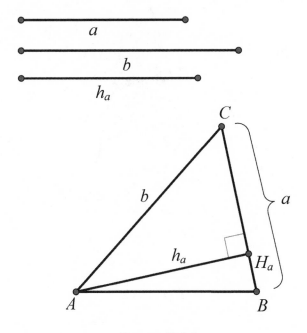

FIGURE 8-14

As we can see in figure 8-14, $\triangle ACH_a$ is a right triangle and is constructible. To do the construction we begin by drawing a circle with center at $C$ and radius length $a$. We show this in figure 8-15. To keep the figure manageable we will not always draw the complete circle, just the arcs that we need for the construction.

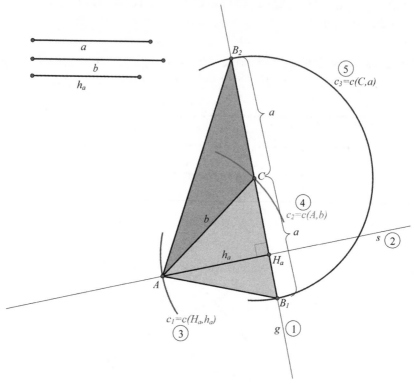

FIGURE 8-15

The steps for the construction are as follows:

① Draw line $g$ with point $H_a$ placed somewhere on line $g$.
② Construct a perpendicular, $s$, to line $g$, at the point $H_a$.
③ Draw circle $c_1 = c(H_a, h_a)$ with center at $H_a$ and radius $h_a$ and label the point of intersection, $A$, with line $s$.
④ Draw circle $c_2 = c(A, b)$ with center at $A$ and radius $b$ to meet line $g$ at point $C$.
⑤ Draw circle $c_3 = c(C, a)$ with center at $C$ and radius $a$ to meet line $g$ at points $B_1$ and $B_2$.

We then have constructed $\triangle AB_1C$ and $\triangle AB_2C$.

Bear in mind that we have here constructed two triangles:

When $b > h_a$, we get two noncongruent triangles.
When $b = h_a$, we get two congruent right triangles.
When $b < h_a$, we are not able to construct a triangle.

## EXAMPLE 4: CONSTRUCT A TRIANGLE ABC, GIVEN (a, b, m_c)

This triangle construction is to produce triangle $ABC$, where two sides ($a$ and $b$) are given and the length of the median ($m_c$) to the third side is also given.

In figure 8-16, we have $a = BC$, $b = AC$, and $m_c = CM_c$. This construction is a bit tricky in that we first have to firm up a triangle within this configuration—one that is not yet there. To do this we will begin by drawing $m_c$ through $M_c$ to point $C'$ so that $CM_c = C'M_c$.

We can now construct triangle $ACC'$, which then allows us to draw $AM_c$. We can then complete the required triangle $ABC$ by extending $AM_c$ to point $B$, where $AM_c = BM_c$. Another possibility is to draw circle $c(C, a)$ and mark the point at which it meets $AM_c$ extended to be the required point $B$.

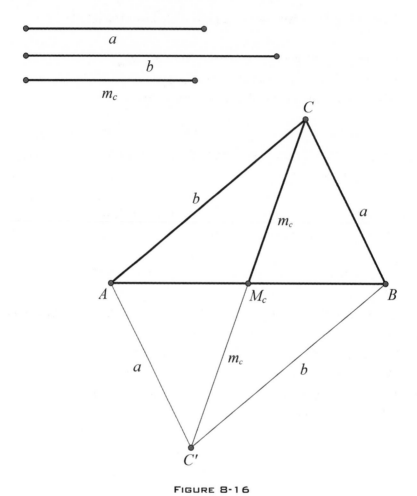

FIGURE 8-16

The steps of the construction (shown in figure 8-17) are as follows:

① Draw line $g$ and mark point $M_c$ on $g$.
② Construct circle $c_1 = c(M_c, m_c)$ with $M_c$ as center and radius $m_c$, intersecting line $g$ at points $C$ and $C'$.
③ Draw circle $c_2 = c(C, b)$ with center $C$ and radius $b$.
④ Draw circle $c_3 = c(C', a)$ with center $C'$ and radius $a$, intersecting circle $c_2 = c(C, b)$ at point $A$.

⑤ Draw circle $c_4 = c(C, a)$ with center $C$ and radius $a$, intersecting $AM_c$ at points $B_1$ and $B_2$.

Our required triangle is then $\triangle AB_1C$, which preserves $m_c$ as the median. (In triangle $AB_2C$, the line $m_c$ is not a median.)

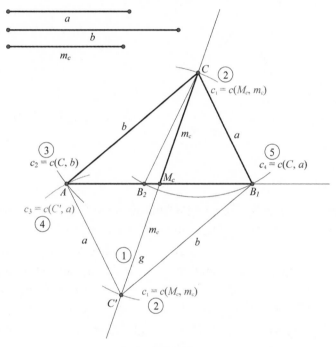

FIGURE 8-17

# EXAMPLE 5: CONSTRUCT TRIANGLE ABC, GIVEN (a, b, t_c)

As with the previous construction, we will need to create a firm triangle from which we can then construct the required triangle. We are given two sides of the triangle that we seek and the bisector of the included angle. That is, we have for triangle $ABC$, $a = BC$, $b = AC$, and $t_c = CT_c$, as shown in figure 8-18.

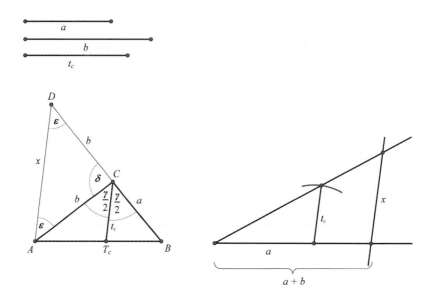

FIGURE 8-18

We shall begin by extending $BC$ the length of $b$, so that we have created isosceles triangle $ACD$, where $AC = CD = b$, $AD = x$, $\delta = \angle ACD$, and $\varepsilon = \angle CAD = \angle CDA$. Both angle $ACT_c$ and angle $CAD$ are one-half the supplement of angle $ACD$. Therefore, they are equal, and $AD$ is then parallel to $CT_c$. Since triangle $ABD$ is similar to triangle $T_cBC$, we get the proportion $\frac{AD}{CT_c} = \frac{BD}{BC}$. Therefore, $\frac{x}{t_c} = \frac{a+b}{a}$. We can then construct $AD = x$, since we know the lengths of $a$, $a + b$, and $t_c$, and the operations of multiplication and division are constructible operations.

We are now ready to do the actual construction. First, we construct triangle $ACD$. Then construct circle $c(C, a)$ with center at $C$ and radius $a$ intersecting $DC$ at point $B$.

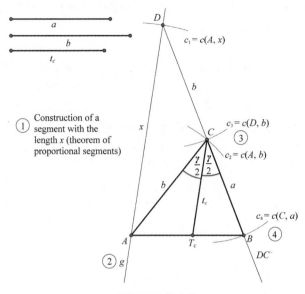

FIGURE 8-19

The construction steps are as follows (figure 8-19):

① Construct $AD = x$ as the result of $x = \dfrac{t_c(a+b)}{a}$.[6]

② Draw line $g$ through points $A$ and $D$, and then measure off the length $x$ along line $g$, with circle $c_1 = c(A, x)$, which will intersect line $g$ at point $D$.

③ We then construct triangle $ACD$ with $AC = CD = b$, $AD = x$ by drawing circle $c_2 = c(A, b)$ and circle $c_3 = c(D, b)$ which intersect at point $C$.

④ Extending $DC$ the length of $a$ from point $C$ by drawing circle $c_4 = c(C, a)$ to intersect $DC$ at point $B$

We have thus constructed triangle $ABC$.

## EXAMPLE 6: TO CONSTRUCT TRIANGLE ABC, GIVEN ($a$, $\angle A$, $h_a$)

Here we must construct a triangle, given the length of a side ($a$), the measure of the angle ($\alpha$) opposite that side, and the altitude ($h_a$) to that given side. In figure 8-20, we see $a = BC$, $\alpha = \angle BAC$, and $h_a = AH_a$.

Using perpendiculars, we will construct a line parallel to $BC$ at a distance of $h_a$ from $BC$. We now focus on triangle $A'CM_a$, where we have side and angles of measure ($90°$, $\frac{a}{2}$, $\delta = 90° - \frac{\alpha}{2}$). The three angles, $\angle BA'C$, $\angle BAC$, and $\angle BA''C$, are all equal, since they are all inscribed in the same arc $\overset{\frown}{BC}$. Triangle $BCA'$ is isosceles with a base angle of $\Delta = \angle BCA'$ $= 90° - \frac{\alpha}{2}$. We are now ready to do the actual construction:

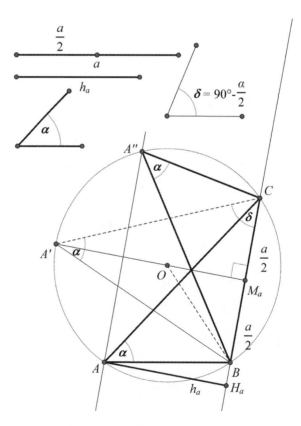

**FIGURE 8-20**

In figure 8-21, we mark the steps of the construction as follows:

① First draw line $g$ and select point $D$ on line $g$.

② Construct a perpendicular $s_1$ to line $g$ at point $D$.

③ Draw circle $c_1 = c(D, h_a)$ with center at $D$ and radius $h_a$ intersecting line $s_1$ at point $E$.

④ Construct perpendicular $s_2$ to $s_1$ at point $E$.

(In short, with steps ② to ④ we simply constructed line $s_2$ parallel to line $g$ at a distance apart of $h_a$.)

⑤ Select point $M_a$ on line $g$ and construct perpendicular $s_3$ to $g$ at point $M_a$.

⑥ Draw circle $c_2 = c\left(M_a, \frac{a}{2}\right)$ with center $M_a$ and radius $\frac{a}{2}$ , intersecting line $g$ at points $B$ and $C$.

⑦ We now need to construct an angle of measure $\delta = 90° - \frac{\alpha}{2}$.

⑧ Now take this angle $\delta$ on line $g$ at point $C$ as shown in figure 8-21.

⑨ Draw the circumscribed circle $c_3$ about triangle $A'BC$ with center $O$. That is, circle $c_3 = c(O, OB)$ with center $O$ and radius $OB$, intersecting line $s_2$ at points $A$ and $A''$.

(Remember, $\angle BA'C$, $\angle BAC$, and $\angle BA''C$ are equal, since they are inscribed in the same arc.)

We then get the required triangle $ABC$.

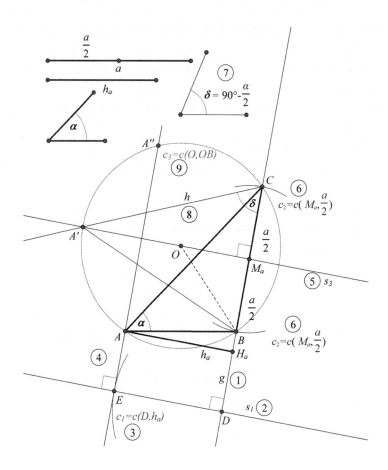

FIGURE 8-21

## EXAMPLE 7: TO CONSTRUCT A TRIANGLE ABC, GIVEN $(a, h_b, h_c)$

Here we are asked to construct a triangle ($\triangle ABC$), given the altitudes ($h_b$ and $h_c$) to two sides and the length of the third side ($a$). In figure 8-22, we show triangle $ABC$ and have the given parts

$$a = BC, h_b = BH_b, \text{ and } h_c = CH_c.$$

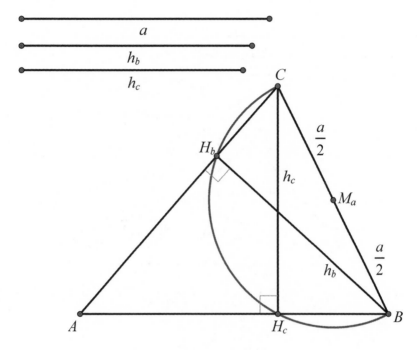

FIGURE 8-22

As we analyze the required construction, we notice that the feet of the two given altitudes must lie on the semicircle with the given side, $a$, as diameter, since an angle inscribed in a semicircle is a right angle. We just need to then mark off the lengths of these altitudes along the semicircle and the remainder is just determining point $A$.

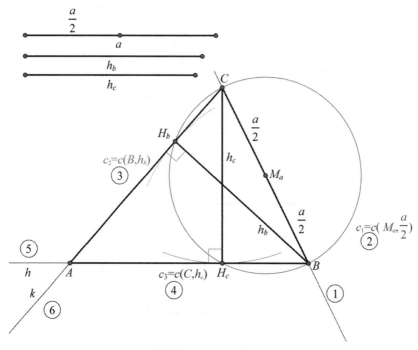

FIGURE 8-23

The construction steps, shown in figure 8-23, are as follows:

① Draw line $g$ and point $M_a$ on line $g$.

② Draw circle $c_1 = c\left(M_a, \frac{a}{2}\right)$ with center at $M_a$ and with radius $\frac{a}{2}$, intersecting the line $g$ at points $B$ and $C$.

③ Draw circle $c_2 = c(B, h_b)$ with center at $B$ and radius $h_b$, intersecting the circle $c_1$ at point $H_b$.

④ Circle $c_3 = c(C, h_c)$ with center at $C$ and radius $h_c$, intersecting the circle $c_1$ at point $H_c$.

⑤ Draw line $h = BH_c$.

⑥ Draw line $k = CH_b$, intersecting $h = BH_c$ at point $A$.

We have, therefore, constructed triangle $ABC$.

(The reader may wish to consider the various cases where $h_b < a$, and $h_c < a$; $h_b = a$, and $h_c < a$; and $h_b < a$, and $h_c = a$.)

## EXAMPLE 8: TO CONSTRUCT A TRIANGLE ABC, GIVEN (a, $h_b$, $m_c$)

This construction requires a triangle, where we are given the length of one side, the length of the altitude to a second side, and the length of the median to the third. In figure 8-24, for triangle $ABC$, we, therefore, have given the line segments $a = BC$, $h_b = BH_b$, and $m_c = CM_c$.

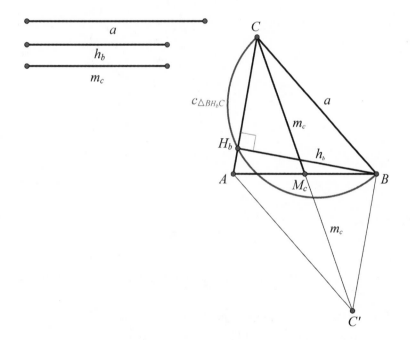

FIGURE 8-24

A quick inspection of figure 8-24 reveals that we can immediately construct right triangle $BCH_b$. We also notice that by extending $CM_c$ its own length to point $C'$, which is where the line containing point $B$ and parallel to $CH_b$ intersect, we can produce a parallelogram $ACBC'$, since $AM_c = BM_c = \frac{c}{2}$, thereby, establishing a quadrilateral, where the diagonals bisect each other. This is done by drawing $BM_c$ to meet $CH_b$ at point $A$.

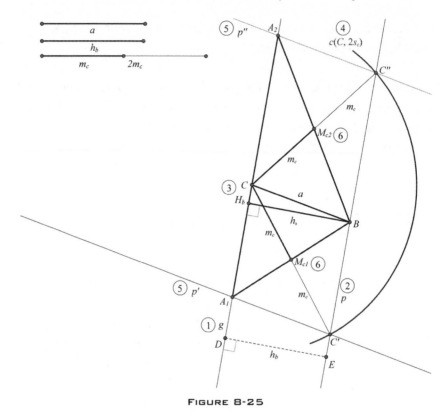

**FIGURE 8-25**

The steps of the construction (shown in figure 8-25) are as follows:

① Draw line $g$ with point $D$ on $g$.

② Construct a line $p$ parallel to line $g$ at a distance $h_b$ apart.

③ Construct the right triangle $BCH_b$, where $a = BC$, $h_b = BH_b$, and $\angle BH_bC = 90°$.

④ Draw circle $c(C, 2m_c)$ with center at $C$ and with radius $2m_c$, intersecting line $p$ at points $C'$ and $C''$.

⑤ Construct line $p'$ parallel to $BC$ at $C'$ intersecting line $g$ at point

$A_1$. (Construct line $p''$ parallel to $BC$ at $C''$ intersecting line $g$ at point $A_2$.)

⑥ Draw line $A_1B$ intersecting $CC'$ at point $M_{c1}$.

(Draw line $A_2B$ intersecting $CC''$ at point $M_{c2}$.)

The result is that we have constructed two triangles: $\Delta A_1BC$ and $\Delta A_2BC$.

## EXAMPLE 9: TO CONSTRUCT TRIANGLE ABC, GIVEN (a, $h_b$, $t_c$)

The construction of triangle $ABC$ gives us one side length ($a$), the length of an altitude ($h_b$) to a second side, and the bisector of the angle opposite the third side of the triangle. In figure 8-26, we have triangle $ABC$ as well as $a = BC$, $h_b = BH_b$, and $t_c = CT_c$.

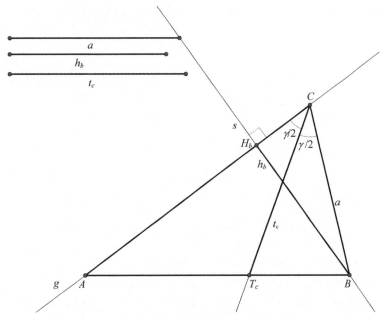

FIGURE 8-26

We begin by constructing right triangle $BCH_b$ (figure 8-26), which can be easily done since we have the length of the hypotenuse and one leg of the right triangle. We then bisect the angle $ACB$ and measure the length $t_c$ along this angle bisector to determine point $T_c$. (Here, when we say "measure" we mean to draw circle $c(C, t_c)$.) This then enables us to complete the triangle construction by drawing $BT_c$ to intersect $CH_b$ at $A$. This constructs one triangle. But as you will see from the construction steps below, there is a second triangle that can be constructed with the given information: the one where the point $C$ is between points $A$ and $H_b$.

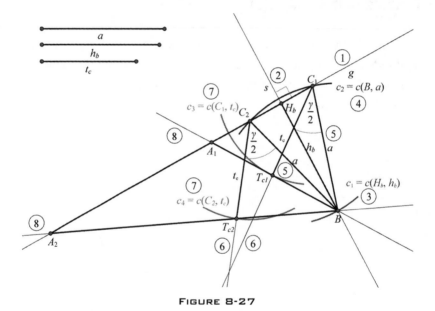

FIGURE 8-27

The construction steps shown in figure 8-27 are as follows:

① Draw line $g$ with point $H_b$ on line $g$.

② Construct a perpendicular line $s$ to line $g$ at point $H_b$.

③ Draw circle $c_1 = c(H_b, h_b)$ with center at $H_b$ and radius $h_b$, intersecting $s$ at point $B$.

④ Draw circle $c_2 = c(B, a)$ with center at $B$ and radius $a$, intersecting line $g$ at points $C_1$ and $C_2$.

⑤ We have sides $BC_1$ (= $a$) and $BC_2$ (= $a$), whereupon we have constructed $\angle H_bC_1B$ (= $\gamma$) and $\angle H_bC_2B$ (= $\gamma$).

⑥ We now construct the bisectors of angles $\angle H_bC_1B$ (= $\gamma$) and $\angle H_bC_2B$
(= $\gamma$).

⑦ Draw circle $c_3 = c(C_1, t_c)$ with center at $C_1$ and radius $t_c$, intersecting
the angle bisector at point $T_{c1}$. Then draw circle $c_4 = c(C_2, t_c)$ with
center at $C_2$ and radius $t_c$, intersecting the angle bisector at point $T_{c2}$.

⑧ Draw the line connecting points $B$ and $T_{c1}$, as well as the line connecting points $B$ and $T_{c2}$; each intersects the line $g$ at points $A_1$ and
$A_2$, respectively.

The result is that we have constructed two triangles from the given
information: $\Delta A_1BC_1$ and $\Delta A_2BC_2$.

You may wish to consider when there are no solutions, or when there
is only one solution.


## EXAMPLE 10: CONSTRUCT TRIANGLE *ABC*, GIVEN (*a*, ∠*C*, *m$_c$*)

For this triangle construction, we are given an angle ($\gamma$) of triangle $ABC$,
one of its adjacent sides ($a$), and the median ($m_c$) to the side opposite the
given angle. In figure 8-28, we have for triangle $ABC$ the following parts
given: $a = BC$, $\gamma = \angle ACB$, and $m_c = CM_c$.

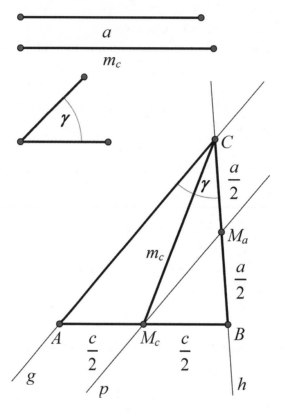

FIGURE 8-28

After we mark off the length of the given side along one of the rays of the given angle, we will locate the midpoint of the given side ($a$) and construct a parallel line to the other ray of the given angle. This will determine the point ($M_c$) along the set of points that the median length follows. (See figure 8-28.) The reason for this is that if a line contains the midpoint of one side of a triangle and is parallel to a second side, then it will also bisect the third side of the triangle.

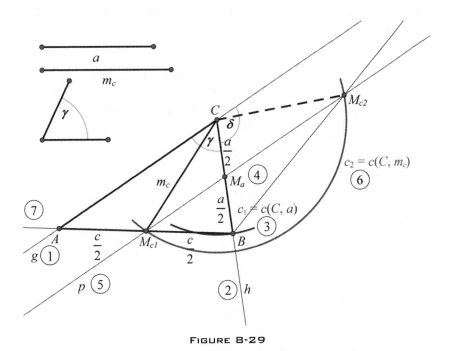

FIGURE 8-29

We now do the construction, as shown in figure 8-29.

① Begin by drawing line $g$ and place point $C$ on line $g$.

② Construct angle $\gamma$ at point $C$ with one of its rays along line $g$ and the other ray we will label $h$.

③ Draw circle $c_1 = c(C, a)$ with center at $C$ and radius $a$, intersecting $h$ at point $B$.

④ Locate the midpoint of $BC$ and label it $M_a$.

⑤ Construct line $p$ parallel to line $g$ and containing the point $M_a$.

⑥ Draw circle $c_2 = c(C, m_c)$ with center at $C$ and radius $m_c$, intersecting $p$ at point $M_c$.

⑦ We now have line $BM_c$ intersecting line $g$ at point $A$.

We have then completed the construction of triangle $ABC$.

## EXAMPLE 11: CONSTRUCT TRIANGLE *ABC*, GIVEN (∠A, ∠B, $t_b$)

This construction requires us to produce triangle *ABC*, given two of its angles ($\alpha$ and $\beta$) and the length of the bisector ($t_b$) of one of them. In figure 8-30, we show the triangle *ABC* to be constructed when we are given $\alpha = \angle BAC$, $\beta = \angle ABC$, and $t_b = BT_b$.

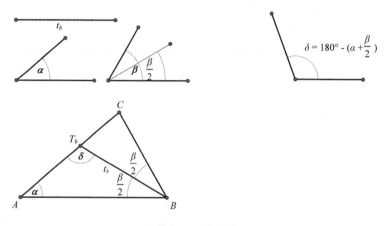

FIGURE 8-30

We will first focus on the two angles of measure $\alpha$ and $\frac{\beta}{2}$, and the length of the angle bisector $t_b$. The third angle of the triangle is $\delta = 180° - (\alpha + \frac{\beta}{2})$. This triangle, $\triangle ABT_b$, is constructible, and it is where we begin our construction of triangle *ABC*.

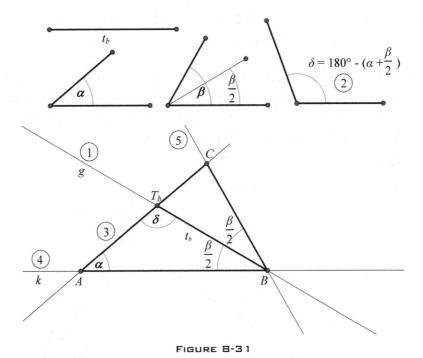

FIGURE 8-31

The construction steps are as follows:

① Draw line $g$, and mark off $t_b = BT_b$ along the line $g$ as shown in figure 8-31.
② We then construct the angle of measure $\delta = 180° - (\alpha + \frac{\beta}{2})$ $= \angle AT_bB$, which we will use as part of the construction.
③ We now take this angle $\delta = 180° - (\alpha + \frac{\beta}{2}) = \angle AT_bB$ at point $T_b$ along line $g$ and create line $h$.
④ At point $B$ construct the angle of measure $\frac{\beta}{2}$ using line $g$ for one ray, and determine line $k$ that will intersect line $h$ at point $A$.
⑤ Then place angle $\beta = \angle ABC$ at point $B$, and have line $l$ intersect line $g$ at point $C$. One can also construct the angle $\frac{\beta}{2}$ at point $B$ to determine line $l$, and then, as before determine point $C$.

We have then completed the construction of triangle $ABC$.

## EXAMPLE 12: TO CONSTRUCT TRIANGLE ABC, GIVEN (∠A, $h_b$, $h_c$)

We are to construct a triangle ($\triangle ABC$) where we are given one of its angles ($\alpha$) and two of its altitudes ($h_b$, $h_c$) not emanating from the vertex of the given angle. In figure 8-32, we see triangle $ABC$ that we are required to construct as well as the given parts $\alpha = \angle BAC$, $h_b = BH_b$ and $h_c = CH_c$.

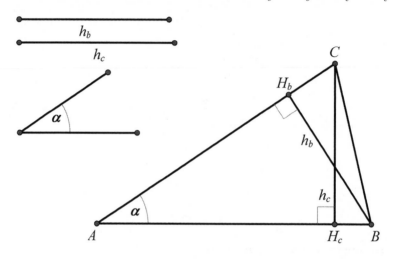

FIGURE 8-32

We begin the triangle construction by drawing the angle $\alpha$, and then finding the set of points that would include the vertex point of each of the two given altitudes. We will do this by constructing the parallel lines that contain these points.

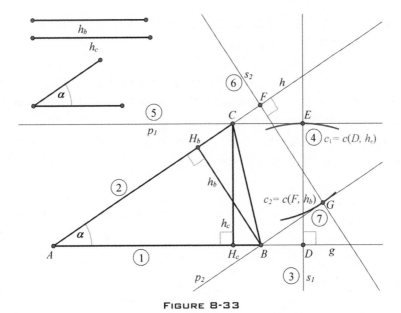

FIGURE 8-33

The steps of the construction are shown in figure 8-33 and are as follows:

① Draw line $g$ and place point $A$ on $g$.

② Construct the angle $\alpha$ at $A$, using line $g$ and determining line $h$.

③ At some convenient point $D$ on line $g$, construct a perpendicular to line $g$—call it $s_1$.

④ Draw circle $c_1 = c(D, h_c)$ with center at $D$ and with radius $h_c$, intersecting $s_1$ at $E$.

⑤ Construct a perpendicular, $p_1$, to line $s_1$ at the point $E$ intersecting $h$ at $C$.

⑥ At some convenient point $F$ on line $h$, construct the perpendicular $s_2$ to line $h$.

⑦ Draw circle $c_2 = c(F, h_b)$ with center at point $F$ and radius $h_b$, intersecting $s_2$ at point $G$.

⑧ Construct a perpendicular, $p_2$, to line $s_2$ at point $G$, intersecting line $g$ at point $B$.

This completes the construction of triangle $ABC$.

## EXAMPLE 13: TO CONSTRUCT TRIANGLE ABC, GIVEN (∠A, $h_b$, $m_a$)

We are to construct a triangle ($\triangle ABC$) when given one of its angles ($\alpha$), the median ($m_a$) drawn from the vertex of that given angle, and the altitude ($h_b$) to another side of the triangle.

In figure 8-34, we show the required triangle, $\triangle ABC$, and the given parts $\alpha = \angle BAC$, $h_b = BH_b$, and $m_a = AM_a$.

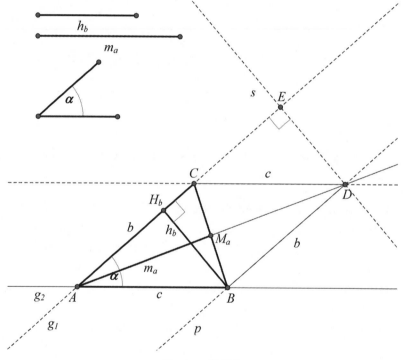

**FIGURE 8-34**

Our plan for this construction is to first consider the doubling of the median $m_a$ so that we can create a parallelogram with side lengths $b$ and $c$, as shown in figure 8-34. The sides of this parallelogram—parallel to the rays of the given angle—one of which will be a distance $h_b$ from the given angle's ray, will deliver the sought-after triangle $ABC$.

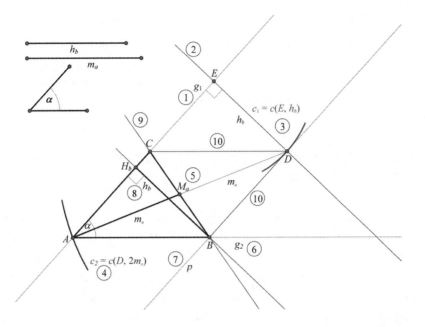

FIGURE 8-35

After this overall strategy for the construction, the following steps shown in figure 8-35 will detail the construction.

① Draw line $g_1$ and place point $E$ on line $g_1$.

② Construct perpendicular line, $s$, to line $g_1$ at point $E$.

③ Draw circle $c_1 = c(E, h_b)$ with center $E$ and radius $h_b$, intersecting line $s$ at point $D$.

④ Draw circle $c_2 = c(D, 2m_a)$ with center $D$ and radius $2m_a$, intersecting line $g_1$ at point $A$.

⑤ Locate the midpoint $M_a$ of segment $AD$ so that $AM_a = DM_a = m_a$.

⑥ At point $A$ construct angle $\alpha$ on line $g_1$. (See figure 8-35 for placement.)

⑦ At point $D$ construct a parallel line, $p$, to line $g_1$, intersecting line $g_2$ at point $B$.

⑧ From point $B$, construct a perpendicular to line $g_1$ and mark off $BH_b$ ($= h_b$) to determine $H_b$.

⑨ We then have $BM_a$ intersecting the line $g_1$ at point $C$.

⑩ The line segments $BD$ and $CD$ complete the parallelogram $ABDC$.

The triangle $ABC$ is then constructed. You may wish to consider when there are no solutions, or when there is only one solution.

## EXAMPLE 14: TO CONSTRUCT TRIANGLE ABC, GIVEN ($h_a$, $h_b$, $h_c$)

We are asked to construct a triangle given the lengths of the three altitudes. In figure 8-36, we are given the lengths of the three altitudes: $h_a = AH_a$, $h_b = BH_b$, and $h_c = CH_c$.

The technique here will be quite different from other triangle constructions. We will seek to construct the triangle composed of the altitude lengths and then convert that to a similar triangle, which would be the required one.

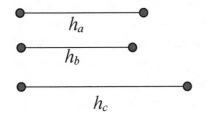

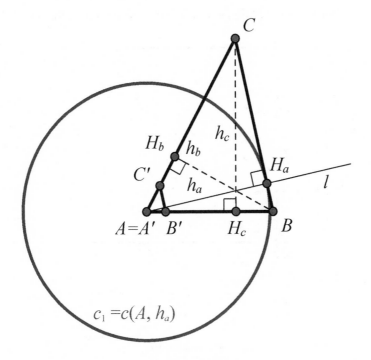

FIGURE 8-36

From the formula for the area of a triangle, we get $Area \; \Delta ABC = \frac{ah_a}{2} = \frac{bh_b}{2} = \frac{ch_c}{2}$, which implies that $ah_a = bh_b = ch_c$, or that $a : \frac{1}{h_a} = b : \frac{1}{h_b} = c : \frac{1}{h_c}$.

We then have $h_a : \frac{1}{a} = h_b : \frac{1}{b} = h_c : \frac{1}{c}$. This shows that the ratio of the sides of a triangle are the reciprocals of the ratio of its altitudes. Therefore, $a{:}b{:}c = \frac{1}{h_a} : \frac{1}{h_b} : \frac{1}{h_c}$, or $h_a{:}h_b{:}h_c = \frac{1}{a} : \frac{1}{b} : \frac{1}{c}$.

Therefore, it follows that every triangle is similar to the triangle composed of the reciprocals of lengths of the altitudes of the orginal triangle.

So triangle $ABC$ is similar to triangle $A'B'C'$, whose sides have lengths $a' = B'C' = \frac{1}{h_a}$, $b' = A'C' = \frac{1}{h_b}$, and $c' = A'B' = \frac{1}{h_c}$. Thus, we will construct the triangle $A'B'C'$. The perpendicular $l$ from point $A = A'$ to $B'C'$ will intersect the circle $c_1 = c(A, h_a)$ with center at $A$ and radius $h_a$ at point $H_a$ on $BC$. The perpendicular line drawn to line $l$ at point $H_a$ intersects the lines $A'B'$ and $A'C'$ at points $B$ and $C$, respectively.

We first have to do some preparatory construction. We need to understand how to construct a segment of length $\frac{1}{b}$, when given the length $b$.

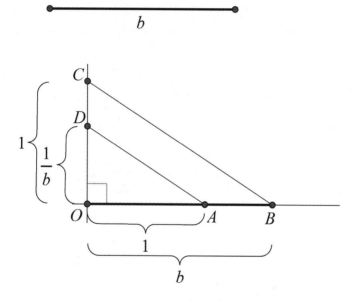

FIGURE 8-37

We draw $OB = b$ and construct $OC = 1$, so that the two segments are perpendicular to each other, as shown in figure 8-37. We mark off $OA = 1$, and construct $AD$ parallel to $BC$. It then follows from triangle similarity that $OD = \frac{OD}{1} = \frac{OD}{OC} = \frac{OA}{OB} = \frac{1}{b}$, or simply put, $OD = \frac{1}{b}$.

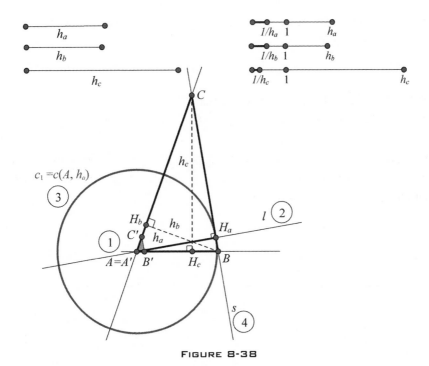

FIGURE 8-38

Now that we have a method for constructing the reciprocal of a given length, we can do the actual construction by first constructing a triangle whose side lengths are the reciprocals of the lengths of the given altitudes. We then construct a triangle similar to this one, but with altitudes of the given lengths.

The following steps refer to figure 8-38.

① Construct a triangle $\triangle A'B'C'$ with side lengths as follows:
$a' = B'C' = \frac{1}{h_a}, b' = A'C' = \frac{1}{h_b}, c' = A'B' = \frac{1}{h_c}$.
(We do this using the above construction for $\frac{1}{b}$ in figure 8-37.)

② Construct a perpendicular line, $l$, from $A = A'$ to $B'C'$.

③ Draw circle $c_1 = c(A, h_a)$ with center $A$ and with radius $h_a$, intersecting $l$ at point $H_a$ on $BC$.

④ Construct the perpendicular line $s$ to the line $l$ at point $H_a$, intersecting line $A'B'$ at point $B$ and intersecting the line $A'C'$ at point $C$.

This then completes the construction of triangle *ABC*, when we were given only its three altitudes.

## EXAMPLE 15: CONSTRUCT A TRIANGLE *ABC*, GIVEN ($h_a$, $h_b$, $m_a$)

In figure 8-39, we are shown the given parts of triangle *ABC*, namely, the two altitudes ($h_a$, $h_b$) and the median ($m_a$) from one of the vertices from which one of the given altitudes was drawn.

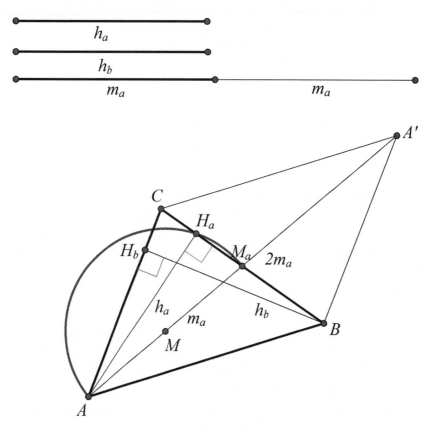

FIGURE 8-39

Thus we have $h_a = AH_a$, $h_b = BH_b$, and $m_a = AM_a$. We would begin by constructing a line parallel to $AC$ at a distance of $h_b$ from it, which will contain point $B$ and will intersect the circle with center $A$ and radius $2m_a$ at point $A'$. We can then construct parallelogram $ACA'B$. We then draw a semicircle on diameter $AM_a = m_a$, intersecting the side $BC$ at the foot $H_a$ of the given altitude $h_a$. This lets us draw the line that will determine $BC$ when it intersects with the two original parallels.

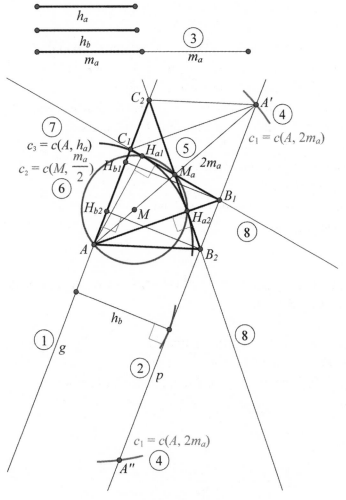

FIGURE 8-40

We shall now do the actual construction in steps below, as shown in figure 8-40:

① Draw line $g$ with point $A$ on line $g$.

② Draw line $p$ parallel to line $g$ at a distance of $h_b$ apart.

③ Construct the length $2m_a$.

④ Draw circle $c_1 = c(A, 2m_a)$ with center $A$ and radius $2m_a$, intersecting $p$ in points $A'$ and $A''$.

⑤ Bisect the line segment $AA'$ ($= 2m_a$) and call the point $M_a$, and bisect $AM_a$ ($= m_a$) to get point $M$.

⑥ Draw the circle $c_2 = c(M, \frac{m_a}{2})$ with center $M$ and radius $\frac{m_a}{2}$.

⑦ Draw circle $c_3 = c(A, h_a)$ with center $A$ and with radius $h_a$, intersecting circle $c_2$ in the points $H_{a1}$ and $H_{a2}$.

⑧ Draw line $M_a H_{a1}$ intersecting line $p$ in $B_1$ and line $g$ in $C_1$; also draw line $M_a H_{a2}$, intersecting line $p$ in $B_2$ and line $g$ in $C_2$.

The result is we have constructed two triangles, $\triangle AB_1C_1$ and $\triangle AB_2C_2$, from the given information.

You may wish to consider when there are no solutions, or when there is only one solution.

## EXAMPLE 16: CONSTRUCT THE TRIANGLE ABC, GIVEN ($h_a$, $m_a$, $t_a$)

We are asked here to construct a triangle, $\triangle ABC$, when we are given the altitude, median, and angle bisector from the same vertex of the triangle ($h_a$, $m_a$, $t_a$). In figure 8-41, we see triangle $ABC$, which we are to construct, given the following parts: $h_a = AH_a$, $m_a = AM_a$, and $t_a = AT_a$.

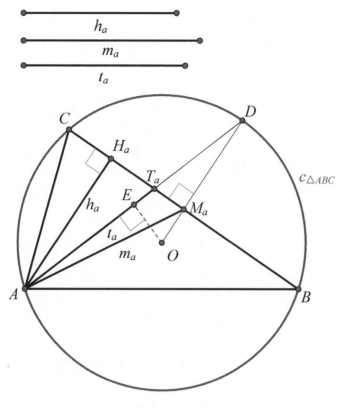

FIGURE 8-41

There are two possibilities: either $h_a < t_a < m_a$, or $h_a = m_a = t_a$. In the latter case, we would have an isosceles triangle and the construction would be relatively trivial. If the three given segments are not equal (i.e., $h_a \neq m_a \neq t_a$), then we would begin the task by constructing the right triangle $AH_aM_a$. Following that, we extend $AT_a$ and the perpendicular to $H_aM_a$ at point $M_a$, which will intersect at point $D$ on the circumscribed circle, as each line bisects arc $BC$. We construct the perpendicular line at $M_a$, and the perpendicular bisector of $AD$—these meet at the center, $O$, of the circumscribed circle. We now draw the circle with center at $A$ and radius $OA$. This then allows us to get the required points $C$ and $B$.

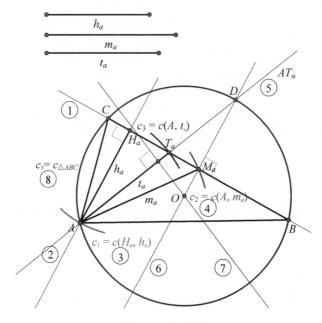

FIGURE 8-42

In figure 8-42, we show the step-by-step construction as follows:

① Draw line $g$ with point $H_a$ on line $g$.

② Construct line $s_1$ perpendicular to line $g$ at point $H_a$.

③ Draw circle $c_1 = c(H_a, h_a)$ with center at $H_a$ and radius $h_a$, intersecting line $s$ at point $A$.

④ Draw circle $c_2 = c(A, m_a)$ with center at $A$ with radius $m_a$, intersecting line $g$ at point $M_a$.

⑤ Draw line $AT_a = t_a$.

⑥ Construct line $s_2$ perpendicular to $H_a M_a$ at point $M_a$ intersecting $AT_a$ at point $D$.

⑦ Construct $s_3$, the perpendicular bisector of $AD$, and intersecting $s_2$ at the center, $O$, of the circumscribed circle.

⑧ Draw the circumscribed circle $c_3 = c_{\triangle ABC} = c(O, OA)$ with center at $O$ and radius $OA$, and intersecting line $g$ at points $B$ and $C$.

We have, thus, completed the construction of triangle *ABC*.

What would result when $h_a = m_a = t_a$? You may wish to consider the case when a triangle *ABC* doesn't exist.

## EXAMPLE 17: CONSTRUCT THE TRIANGLE ABC, GIVEN ($m_a$, $m_b$, $m_c$)

Here we are required to construct a triangle *ABC*, where the three medians are given (figure 8-43). Symbolically we have the following: $m_a = AM_a$, $m_b = BM_b$, and $m_c = CM_c$. The aim of this construction is to first construct triangle *BGK*, which is composed of sides, each of which is $\frac{2}{3}$ the length of one of the given medians.

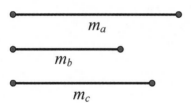

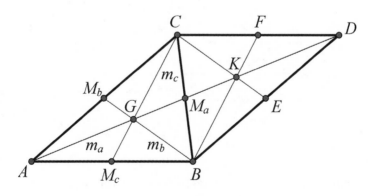

FIGURE 8-43

We know the centroid of a triangle divides each of the medians at a trisection point. Therefore,

$$BG = \frac{2}{3}BM_b = \frac{2}{3}m_b,$$

$$BK = \frac{2}{3}CM_c = \frac{2}{3}m_c, \text{ and}$$

$$GK = GM_a + M_aK = \frac{1}{3}AM_a + \frac{1}{3}AM_a = 2 \cdot \frac{1}{3}AM_a = \frac{2}{3}m_a.$$

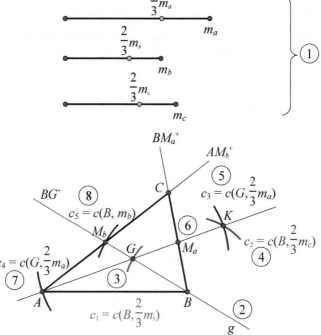

FIGURE 8-44

The construction pictured in figure 8-44 is as follows:

① We first trisect each of the given median lengths, and then we will use:

$$\frac{2}{3}m_b, \frac{2}{3}m_c, \text{ and } \frac{2}{3}m_a.$$

② Draw line $g$ with point $B$ on line $g$.

③ Draw circle $c_1 = c(B, \frac{2}{3}m_b)$ with center $B$ and radius $\frac{2}{3}m_b$, intersecting line $g$ at point $G$.

④ Draw circle $c_2 = c(B, \frac{2}{32}m_c)$ with center $B$ and radius $\frac{2}{3}m_c$.

⑤ Draw circle $c_3 = c(G, \frac{2}{3}m_a)$ with center $B$ and radius $\frac{2}{3}m_a$, intersecting circle $c_2$ at point $K$.

⑥ Draw line segment $GK$ to bisect $CB$ at midpoint $M_a$.

⑦ Draw circle $c_4 = c(G, \frac{2}{3}m_a)$ with center at $G$ and radius $\frac{2}{3}m_a$, intersecting line $GK$ at point $A$.

⑧ Draw circle $c_5 = c(B, m_b)$ with center at $B$ and radius $m_b$, intersecting $BG$ at point $M_b$.

⑨ Draw $AM_b$ to intersect $BM_a$ at point $C$.

The result is the sought-after triangle $ABC$.

What happens when $m_a = m_b$ and $m_c < 2\,m_a$? The reader may also wish to consider the case when a triangle $ABC$ doesn't exist.

At this point one would expect to be asked to construct a triangle $ABC$, given the lengths of the three angle bisectors $(t_a, t_b, t_c)$. However, this is not a constructible triangle in the general case using only an unmarked straightedge and a pair of compasses. For some special situations it would be possible, but we are concerned here with general cases.

## EXAMPLE 18: CONSTRUCT A TRIANGLE ABC, GIVEN (a, $m_b$, R)

We are asked here to construct a triangle $ABC$, where we are given one side $(a)$, the median to another side $(m_b)$, and the radius $(R)$ of the triangle's circumscribed circle. In figure 8-45, we see the triangle to be constructed, $\triangle ABC$, and the given parts $a = BC$, $m_b = BM_b$, and $R = OA = OB = OC$.

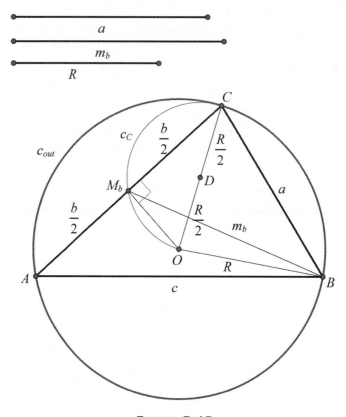

FIGURE 8-45

To do this construction we begin by drawing the given circumscribed circle (with center at $O$) of triangle $ABC$ with radius $R$. The midpoint of side $AC$ is point $M_b$. The semicircle on $OC$ is used to determine the right angle so that the segment from the center of the circumscribed circle and the length $m_b$ meet at the common point $M_b$. Once this point has been determined, the remainder of the construction is simply to draw $CM_b$ to meet the circumscribed circle at point $A$.

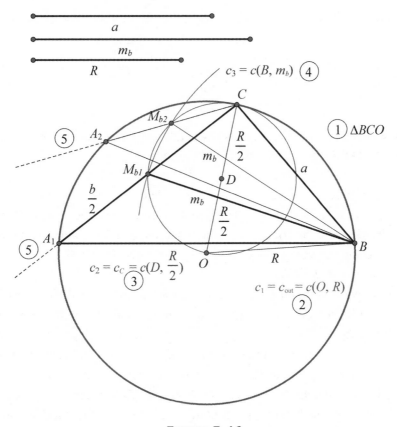

$a$

$m_b$

$R$

FIGURE 8-46

The construction follows the above outline and can be seen in figure 8-46:

① Construct the isosceles triangle $BCO$ with $a = BC$ and $R = BO = CO$.

② Draw circle $c_1 = c(O, R)$ with center at $O$ and radius $R$.

③ Bisect line segment $CO$ to obtain point $D$. Then draw circle $c_2 = c\left(D, \dfrac{R}{2}\right)$ with center $D$ and with radius $\dfrac{R}{2}$.

④ Draw circle $c_3 = c(B, m_b)$ with center $B$ and radius $m_b$, intersecting circle $c_2 = c_C$ at points $M_{b1}$ and $M_{b2}$.

⑤ Draw $CM_{b1}$ and $CM_{b2}$ intersecting the circumscribed circle $c_1$ at points $A_1$ and $A_2$, respectively.

We have then constructed two noncongruent triangles: $\Delta A_1BC$ and $\Delta A_2BC$.

Consider the condition for the existence of a solution:

$$\sqrt{R^2 + 2a^2} - R \leq 2m_b \leq \sqrt{R^2 + 2a^2} + R.$$

## EXAMPLE 19: CONSTRUCT A TRIANGLE ABC, GIVEN ($\angle A$, $m_c$, $R$)

We are required to construct a triangle ($\Delta ABC$), where an angle ($\alpha$) is given, as are the median ($m_c$) from another vertex and the radius ($R$) of the triangle's circumscribed circle. In figure 8-47, we have the triangle we seek to construct, $\Delta ABC$, as well as the given parts $\alpha = \angle BAC$, $m_c = CM_c$, and $R = AO = BO = CO$.

Here we have point $O$ as the center of the circumscribed circle of triangle $ABC$. The midpoint of $AB$ is the point $M_c$, and $M_{BO}$ is the midpoint of $BO$. (Therefore, $OM_{BO} = M_{BO}B = \frac{R}{2}$.) Both angles $\delta = \angle BOC$ and $\alpha = \angle BAC$ are inscribed in the same arc, $\overparen{BC}$. Since one is a central angle and the other is an inscribed angle, we know that $\delta = 2\alpha$. We then have the isosceles triangle $BCO$, where $R = BO$, $\Delta = \angle BOC = 2\alpha$, and $R = CO$.

The perpendicular from point $O$ must pass through the midpoint, $M_c$, of side $c = AB$. The point $M_c$ must therefore lie on the semicircle with center $M_{BO}$ and radius $\frac{R}{2}$, as does the point $M_c$, which is determined by the circle with center $C$ and radius $m_c$.

The point $A$ is then determined when $BM_c$ intersects the circle with center $O$ and radius $R = BO = CO$.

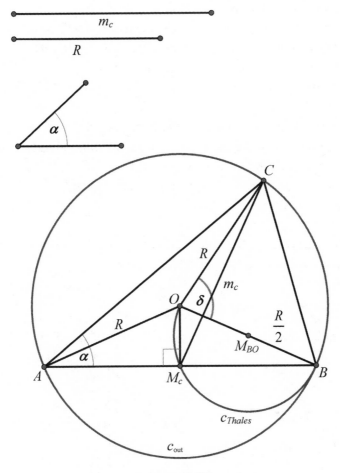

FIGURE 8-47

We shall refer to figure 8-48 in the step-by-step construction of triangle *ABC* as follows:

①  Draw line *g* with point *O* on line *g*.
②  Construct the angle $\delta = 2\alpha$ along line *g* at point *O*—with the other ray of the angle called *h*.
③  Draw circle $c_1 = c(O, R)$ with center *O* and radius *R*, intersecting lines *g* and *h* at points *B* and *C*, respectively.

④ Draw $BO$ and bisect it at $M_{BO}$.

⑤ Draw circle $c_2 = c(M_{BO}, \frac{R}{2})$ with center $M_{BO}$ and radius $\frac{R}{2}$.

⑥ Draw circle $c_3 = c(C, m_c)$ with center $C$ and radius $m_c$, intersecting the circle $c_1 = c(O, R)$ at points $M_{c1}$ and $M_{c2}$.

⑦ Draw $BM_{c1}$ intersecting the circle $c_1 = c(O, R)$ at the point $A_1$.

Similarly, we draw $BM_{c2}$ to intersect circle $c_1 = c(O, R)$ at point $A_2$.

The result is that we have constructed two noncongruent triangles, $\triangle A_1BC$ and $\triangle A_2BC$, from the given parts.

You may wish to consider when there are no solutions, or if there is only one solution.

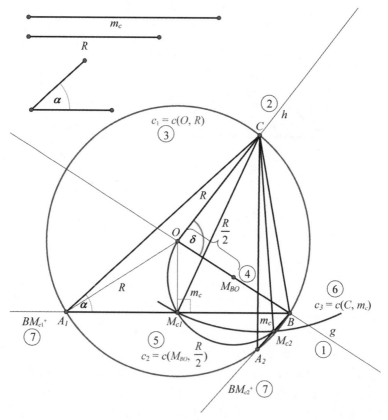

FIGURE 8-48

There are many more such possible triangle constructions. We have merely selected a few that would exhibit the various techniques needed to carry out many triangle constructions. Some of the remaining constructions are quite difficult and will provide a worthy challenge. Others may be rather simple. We encourage the reader to delve into some of the remaining triangle constructions—mindful that this is one of the truest forms of problem solving in the field of Euclidean geometry.

# CHAPTER 9

# INEQUALITIES IN A TRIANGLE

To this point we have concentrated on the relative sizes and areas of triangles and on the symmetric aspects of triangles—of which collinearity, concurrency, symmetry, and certainly equalities are integral parts. We will now explore another aspect of triangles: the inequalities that can be found in and about triangles.

We began our introduction to triangles in chapter 1 by considering what must be true about the relative lengths of the sides of a triangle in order for it to even be a triangle, namely, that the sum of the lengths of any two sides must be greater than the length of the third side. We can see this again in figure 9-1, where it appears that the sum $a + b < c$.

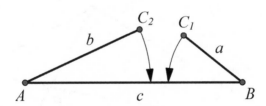

FIGURE 9-1

This means for the triangle $ABC$ to exist in figure 9-2, we have the following inequalities:

$$a + b < c,$$
$$a + c < b,$$
$$b + c < a.$$

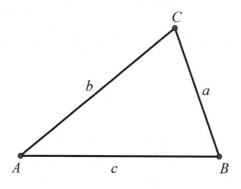

**FIGURE 9-2**

This *triangle inequality* allows us to establish many interesting inequalities in geometry. For example, we can show that the sum of the diagonals of any convex quadrilateral is greater than the sum of either pair of opposite sides.

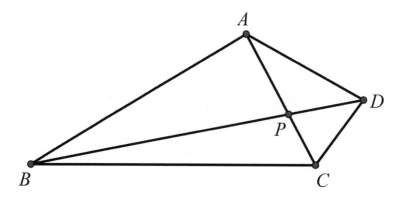

**FIGURE 9-3**

Applying the triangle inequality to triangles *APB* and *CPD*, we get

$$AP + BP > AB$$
$$DP + CP > CD.$$

By addition we get $AP + BP + DP + CP = (AP + CP) + (BP + DP)$ $= AC + BD > AB + CD$.

Therefore, the sum of the diagonals has been shown to be greater than the sum of one pair of opposite sides. That is, $AC + BD > AB + CD$. The same can be shown for the other pair of opposite sides.

## INEQUALITIES BETWEEN THE SIDES

Suppose we add $a + b$ to both sides of the triangle inequality $a + b > c$. We get $2a + 2b > a + b + c$, which can be rewritten as $a + b > \frac{1}{2}(a + b + c)$.

Analogously, using the other sides of a triangle, we can state: $a + c > \frac{1}{2}(a + b + c)$ and $b + c > \frac{1}{2}(a + b + c)$.

Now taking the reciprocals of these inequalities we have

$$\frac{1}{a+b} < \frac{2}{a+b+c}, \frac{1}{a+c} < \frac{2}{a+b+c}, \text{ and } \frac{1}{b+c} < \frac{2}{a+b+c}.$$

We now multiply each by a common factor to get

$$\frac{c}{a+b} < \frac{2c}{a+b+c}, \frac{b}{a+c} < \frac{2b}{a+b+c}, \text{ and } \frac{a}{b+c} < \frac{2a}{a+b+c},$$

which by addition of inequalities gives us

$$\frac{a}{b+c} + \frac{b}{a+c} + \frac{c}{a+b} < \frac{2a}{a+b+c} + \frac{2b}{a+b+c} + \frac{2c}{a+b+c} = \frac{2(a+b+c)}{a+b+c} = 2,$$

or $\frac{a}{b+c} + \frac{b}{a+c} + \frac{c}{a+b} < 2$. It can be shown that this constant 2 cannot be reduced.

Suppose we now investigate the product of the sides of a triangle, namely, $a$, $b$, and $c$. We then need to reintroduce the semiperimeter, $s = \frac{1}{2}(a+b+c)$. We will now introduce three variables:

$$x = s - a = \frac{1}{2}(b+c-a), y = s - b = \frac{1}{2}(c+a-b), \text{ and } z = s - c = \frac{1}{2}(a+b-c).$$

By addition we get $a = y + z$, $b = z + x$, and $c = x + y$. As a result of the triangle inequality we will always have $x, y, z > 0$. We may recognize the lengths $x$, $y$, and $z$ as the tangent segments to the inscribed circle of a triangle (see figure 9-4).

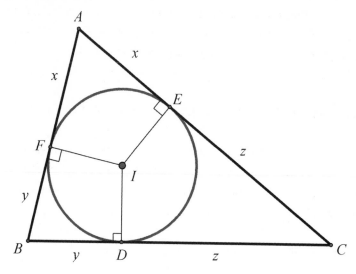

FIGURE 9-4

We can state that $2x + 2y + 2z = a + b + c$, or that $x + y + z = \frac{1}{2}(a + b + c) = s$.

Recall that $a = y + z$, $b = z + x$, and $c = x + y$.

We are now ready to consider the product of the sides $a \cdot b \cdot c$ as follows:

$$a \cdot b \cdot c = (y + z)(z + x)(x + y).$$

The product $(a + b - c)(b + c - a)(c + a - b)$ leads to the following:

$$(a + b - c)(b + c - a)(c + a - b) = 2z \cdot 2x \cdot 2y = 8xyz.$$

The inequality $(y + z)(z + x)(x + y) \geq 2z \cdot 2x \cdot 2y = 8xyz$, first published by Ludolph Lehmus (1780–1863) and later by Alessandro Padoa (1868–1937), stems from the following inequality[1] $\left(\sqrt{x} - \sqrt{y}\right)^2 \geq 0$ and gives us

$$a \cdot b \cdot c \geq (a + b - c)(b + c - a)(c + a - b).$$

We will return to this inequality to prove the famous *Euler inequality*. (See pp. 287–88.)

We can also relate the sum of the squares of the side and the sum of the side as follows:

$$\frac{1}{3} \le \frac{a^2+b^2+c^2}{(a+b+c)^2} \le \frac{1}{2}, \text{ where } \frac{1}{2} \text{ is the lowest bound.}$$

Among the many other inequalities involving the sides of a triangle are

$$\sqrt{a+b-c} + \sqrt{b+c-a} + \sqrt{c+a-b} \ge \sqrt{a} + \sqrt{b} + \sqrt{c},$$

or $a^2b\,(a-b) + b^2c\,(b-c) + c^2a\,(c-a) \ge 0$.

The proofs of these can be found in some of the references we provide later.

## EXTERIOR ANGLES

There are also inequalities involving the angles of a triangle. The most basic of these is that any *exterior angle* of a triangle is greater than either *remote interior angle*. In figure 9-5, angle *ACD* is an exterior angle, and its two *remote* interior angles are angles *ABC* and *BAC*.

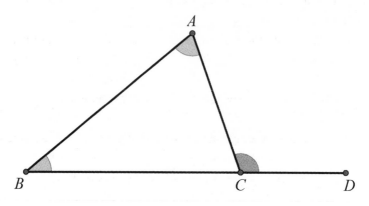

FIGURE 9-5

We can use a clever technique to justify this claim of inequality. We will draw a few lines as shown in figure 9-6. Through the midpoint $M$ of $AC$, we extend $BM$ its own length to point $E$. From the congruent triangles $ABM$ and $CEM$, we have $\angle ACE \cong \angle A$, yet $\angle ACD > \angle ACE$, and, therefore, also $\angle ACD > \angle A$, which is one of the remote interior angles. Because $AB\|EC$, we have $\angle ECD \cong \angle ABC$. But, $\angle ACD > \angle ECD$. Therefore, $\angle ACD > \angle ABC$, which shows that the other remote interior angle ($\angle ABC$) is also less than the exterior angle ($\angle ACD$).

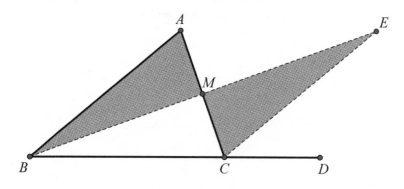

FIGURE 9-6

This relationship will help us prove something that will look obvious, namely, that in figure 9-7 $\angle BDC > \angle A$. To do that, we will extend $BD$ through point $E$, its intersection point with $AC$.

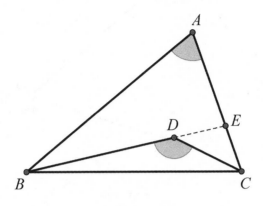

FIGURE 9-7

Considering triangle $DEC$, angle $BDC$ is an exterior angle and angle $DEC$ is a remote interior angle. Therefore, $\angle BDC > \angle DEC$. However, for triangle $ABE$, exterior $\angle DEC > \angle A$. Thus, $\angle BDC > \angle A$.

We can also show some interesting relationships in other geometric figures using this single inequality relationship of a triangle. Consider the quadrilateral $ABCD$ in figure 9-8. Here we can show that $\angle EBC + \angle FDC > \frac{1}{2} (\angle A + \angle C)$. (Here is a hint on how to prove this inequality: Draw $AC$ and consider the angles $EBC$ and $FDC$ as exterior angles of the two newly formed triangles. Then apply the external angle inequality.)

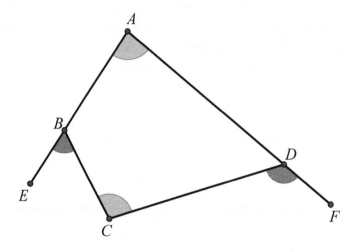

FIGURE 9-8

## FURTHER INEQUALITIES IN A TRIANGLE

We can determine the largest and smallest sides or angles of a triangle if we know one or the other about either the sides or the angles. Sounds confusing? Simply said: If two sides of a triangle are not equal, then the angles opposite them are also not equal—with the greater angle being opposite the greater side. We can show this to be true by using our previously established relationship involving the exterior angle of a triangle.

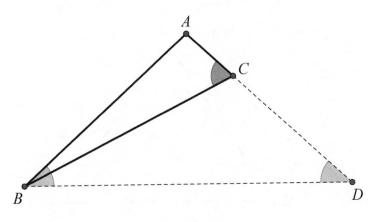

FIGURE 9-9

In figure 9-9, we can clearly see that $AB > AC$. So we will extend $AC$ to point $D$, so that $AB = AD$. Exterior angle $ACB$ is greater than angle $D$. Since $\angle D \cong \angle ABD$, and $\angle ABD > \angle ABC$, it is certainly true that $\angle D > \angle ABC$. However, exterior angle $ACB$ is greater than angle $D$. Therefore, $\angle ACB > \angle ABC$, which we sought to justify.

It is also true that when two angles of a triangle are not equal, the greater side of the triangle is opposite the greater angle.

These relationships are useful, as we can see from the configuration of figure 9-10, where the legs of isosceles triangle $ABC$ are lengthened with $BE > CD$. We can immediately see that since $AE > AB$, in triangle $AED$, $\angle D > \angle E$.

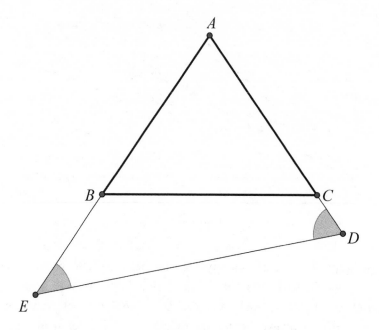

FIGURE 9-10

We can also use this relationship to justify the fact that the shortest distance from a point to a line is the perpendicular distance. To do this we draw any line other than the perpendicular (*PD*) from point *P* to line *AB*; suppose we draw *PC*. (See figure 9-11.) Since $\angle PDC > \angle PCD$, we have $PC > PD$, thus implying *PD* is the shortest such line from point *P* to *AB*.

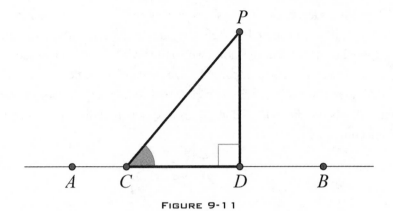

FIGURE 9-11

## USING TRIANGLES TO COMPARE MEANS

The *arithmetic mean* (AM), often simply called "the average," is one of the earliest taught concepts from the field of statistics. It is simply the sum of the items being averaged divided by the number of items. For two items $a$ and $b$ the arithmetic mean (or average) is $\frac{a+b}{2}$.

The *geometric mean* (GM) may be encountered in geometry (often referred to as the mean proportional) and is the $n$th root of the product of the $n$ items. For two items $a$ and $b$ the geometric mean is $\sqrt{ab}$.

A third mean, the *harmonic mean* (HM), is not too popular, since there is not much that can be done with it. It usually comes up as the mean of rates over the same base. The harmonic mean is useful for sets of numbers that are defined in relation to some unit, for example, speed (distance per unit of time).

The harmonic mean of $n$ items is $n$ times the product of the $n$ items divided by the sum of products of the $n$ items taken $(n-1)$ at a time. This may sound a bit difficult to follow, so we offer an equivalent definition as the reciprocal of the arithmetic mean of the reciprocals. Let's see how that manifests itself for a few items.

For two items $a$ and $b$ it is $\dfrac{1}{\frac{\frac{1}{a}+\frac{1}{b}}{2}} = \dfrac{2}{\frac{1}{a}+\frac{1}{b}} = \dfrac{2ab}{a+b}$,

yet for three items, $a$, $b$, and $c$, it is $\dfrac{1}{\frac{\frac{1}{a}+\frac{1}{b}+\frac{1}{c}}{3}} = \dfrac{3}{\frac{1}{a}+\frac{1}{b}+\frac{1}{c}} = \dfrac{3abc}{ab+ac+bc}$.

There are clearly other measures of central tendency, or means, but these will not be considered as we show how triangles exhibit these three more popular means. Furthermore, we shall show how the magnitudes of these three means can be compared in size by using triangles.

Consider a semicircle with center $O$ and radius $OP$. A perpendicular from $P$ meets the diameter $AB$ at $Q$. From $Q$ a perpendicular is drawn to $OP$, meeting it at $S$.

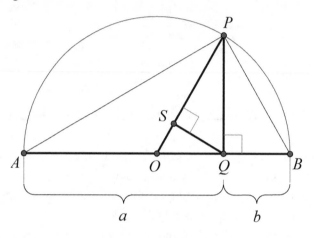

FIGURE 9-12

Let us begin by designating our key line segments, which will be used in our comparisons of means. They are $AQ = a$ and $BQ = b$.

We now have to find line segments in figure 9-12 that represent the various means and represent them in terms of $a$ and $b$.

We can show that the arithmetic mean for $a$ and $b$ is $OP$ with the following justification:

$$OP = OA = OB = \frac{1}{2}AB = \frac{1}{2}(AQ + BQ) = \frac{1}{2}(a + b).$$

The geometric mean for $a$ and $b$ is $PQ$, which is justified as follows. We can show that $\triangle BPQ \sim \triangle APQ$; therefore,

$$\frac{BQ}{PQ} = \frac{PQ}{AQ}, \text{ or } PQ^2 = AQ \cdot BQ = a \cdot b; \text{ so that } PQ = \sqrt{ab}.$$

The harmonic mean for $a$ and $b$ is $PS$, which can be justified as follows.

Now we have $\triangle OPQ \sim \triangle PQS$; therefore, $\dfrac{OP}{PQ} = \dfrac{PQ}{PS}$, or $PS = \dfrac{PQ^2}{OP}$, but $PQ^2 = ab$; and $OP = \dfrac{1}{2}(a+b)$;

therefore, we get $PS = \dfrac{ab}{\dfrac{1}{2}(a+b)} = \dfrac{2ab}{a+b}$.

We are now ready to do the comparisons of these three means. As we consider right triangle $OPQ$, we should recall that the hypotenuse is the longest side of a right triangle. Here this gives us $OP > PQ$. In triangle $PQS$, $PQ > PS$. Therefore, $OP > PQ > PS$.

If $OP \perp AB$, then $O = Q = S$. That means $a = AQ = OA = OB = BQ = b$ and $PQ = PS = OP = a = b$.

All in all, we have $OP \geq PQ \geq PS$. That is, the *arithmetic mean* is greater than or equal to the *geometric mean*, which is greater than or equal to the *harmonic mean*: $AM \geq GM \geq HM$.

## THE MEDIANS OF A TRIANGLE REVISITED

Let us now focus on the sum of the medians of a triangle. First, we will show that for any triangle, the sum of the lengths of the medians is less than the perimeter of the triangle. In figure 9-13, triangle $ABC$ has medians $AD, BE$, and $CF$. We begin our exploration by choosing $N$ on the extension of $AD$ so that $AD = DN$.

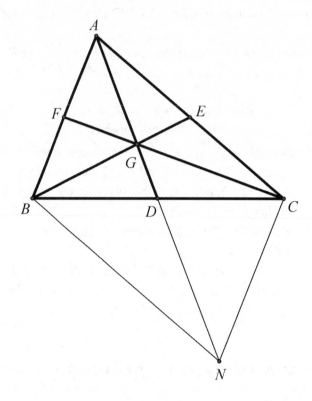

FIGURE 9-13

Quadrilateral $ACNB$ is a parallelogram since the diagonals bisect each other. Therefore $BN = AC$.

For triangle $ABN$, we can state that $AN < AB + BN$, which, with appropriate substitutions, leads us to $2\,AD < AB + AC$, or $2m_a < c + b$.

When we apply this relationship to the other medians, we get
$2m_b < a + c$ and $2m_c < a + b$.

By adding these three inequalities we arrive at $2(m_a + m_b + m_c) < 2(a + b + c)$, or, said more simply, $m_a + m_b + m_c < a + b + c$, which is what we sought to show.

Comparing the sum of the lengths of the medians of a triangle to the triangle's perimeter results in an unexpected relationship, namely, that the sum of the lengths of the medians is greater than three-fourths of the perimeter of the triangle.

To demonstrate this relationship we begin by using the trisection property of the centroid, $G$, of triangle $ABC$.

In triangle $BGC$ (see figure 9-12), we have

$$BG + CG > BC, \text{ or } \frac{2}{3}m_b + \frac{2}{3}m_c > a.$$

Applying this relationship on the other pairs of medians enables us to state the following:

$$\frac{2}{3}m_a + \frac{2}{3}m_c > b, \text{ and } \frac{2}{3}m_a + \frac{2}{3}m_b > c.$$

By adding these three inequalities, we get $\frac{4}{3}(m_a + m_b + m_c) > a+b+c.$

Therefore, $m_a + m_b + m_c > \frac{3}{4}(a+b+c).$

Now combining the two previous median relationships yields the following summary:

$$\frac{3}{4}(a + b + c) < m_a + m_b + m_c < a + b + c.$$

## COMPARING ALTITUDES, ANGLE BISECTORS, AND MEDIANS OF A TRIANGLE

Among the altitudes, the angle bisectors, and the medians of a triangle there are some simple inequalities. Let's consider the relationship between the angle bisector and the median from the same vertex of a triangle. In figure 9-14 we then have $t_a \leq m_a$, $t_b \leq m_b$, and $t_c \leq m_c$.

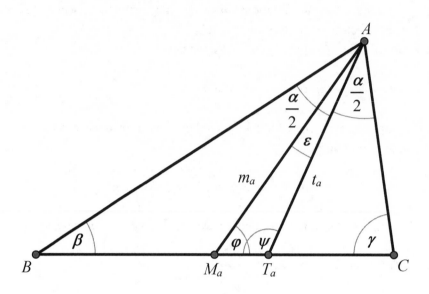

**FIGURE 9-14**

To justify this claim, we will begin with triangle $ABC$ (figure 9-14), where we have $c \geq b$, therefore $\gamma \geq \beta$.

We also have

$\Delta AM_aC$:      $\angle AM_aC + \angle M_aCA + \angle CAM_a = \varphi + \gamma + (\frac{\alpha}{2} + \varepsilon) = 180°$
$\Delta AT_aB$:      $\angle AT_aB + \angle T_aBA + \angle BAT_a = \psi + \beta + \frac{\alpha}{2} = 180°$.

Therefore, $\varphi + \gamma + (\frac{\alpha}{2} + \varepsilon) = \psi + \beta + \frac{\alpha}{2}$, which gives us $\varphi + \gamma + \varepsilon = \psi + \beta$.

This can be rewritten as $\psi = \varphi + \varepsilon + (\gamma - \beta)$.

If $\varepsilon \geq 0$ and $\gamma - \beta \geq 0$, it follows that $\psi \geq \varphi$, which, in triangle $M_aAT_a$, then allows us to conclude that $m_a \geq t_a$, since the greater side is opposite the greater angle.

If we now include the altitudes in this comparison, we get another surprising relationship between these three line segments drawn from the same vertex of any triangle, as we can see in figure 9-15:

$$h_a \leq t_a \leq m_a, h_b \leq t_b \leq m_b, \text{ and } h_c \leq t_c \leq m_c.$$

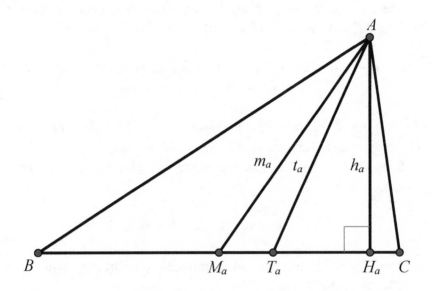

FIGURE 9-15

Beginning with the previously proved inequality, we need only to show that $h_a \leq t_a$, $h_b \leq t_b$, and $h_c \leq t_c$.

We will proceed by showing that $h_a \leq t_a$.

We have altitude $h_a = AH_a$, which is the shortest distance from $A$ to $BC$, which allows us to conclude that $h_a \leq t_a$. Therefore, we established $h_a \leq t_a \leq m_a$. A similar inequality follows for the other vertices.

For the angle-bisector sum, we can also show another nice inequality: $t_a + t_b + t_c \leq \frac{3}{2}(a+b+c)$.

In figure 9-15, we have the following inequalities:

For $\triangle ABT_a$: $AT_a \leq AB + BT_a = c + BT_a$, and

for $\triangle ACT_a$: $AT_a \leq AC + CT_a = b + CT_a$.

Letting $t_a = AT_a$ and by adding the above two statements we get $2t_a \leq c + b + BT_a + CT_a = c + b + a$. It then follows that $t_a \leq \frac{a+b+c}{2}$.

Analogously, we get $t_b \leq \frac{a+b+c}{2}$ and $t_c \leq \frac{a+b+c}{2}$.

Therefore, by addition, $t_a + t_b + t_c \leq \frac{3}{2}(a + b + c)$.

The sum of the angle bisectors can be further refined as Šefket Arslanagić (2010) showed with the following:

$$t_a + t_b + t_c < a + b + c;$$ and, furthermore, we can state that
$$t_a + t_b + t_c \leq \frac{3}{2}(a + b + c).$$

However, we should note that the *equality* of the last relationship only holds true for equilateral triangles, where $a = b = c$ and $t_a = t_b = t_c = \frac{\sqrt{3}}{2}a$.

Surprisingly, we have an analogous inequality involving the altitudes of a triangle.

$$h_a + h_b + h_c \leq \frac{\sqrt{3}}{2}(a + b + c).$$

There are still further unexpected inequalities relating the altitudes to the sides of a triangle, such as

$$\frac{a^2}{h_b h_c} + \frac{b^2}{h_a h_c} + \frac{c^2}{h_a h_b} \geq 4.$$

We can even take this a step further in relating the sum of the squares of the altitudes to the sides:

$$h_a^2 + h_b^2 + h_c^2 > \frac{3}{4}(a^2 + b^2 + c^2).$$

Again, the *equality* in the last statement only holds for equilateral triangles, which can be proved using Heron's formula for the area of a triangle. (See chapter 7, p. 168.)

Finally, we have the following inequality involving altitudes and medians:

$$\frac{h_a}{m_b} + \frac{h_b}{m_c} + \frac{h_c}{m_a} \leq 3.$$

## INEQUALITIES INVOLVING
## A RANDOM POINT IN A TRIANGLE

Suppose we select a random point, $P$, in triangle $ABC$. With what we have so far established about triangle inequalities, we can show in figure 9-16 that

$$PA + PB + PB > \frac{1}{2}(AB + AC + BC).$$

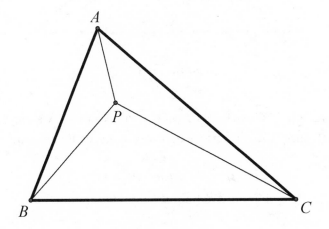

FIGURE 9-16

This can be easily done by applying the triangle inequality to each of the three triangles inside triangle $ABC$ as follows:

$$PA + PB > AB$$
$$PA + PC > AC$$
$$PB + PC > BC$$

Therefore, by addition, $2(PA + PB + PC) > AB + AC + BC$, or, written another way, $PA + PB + PB > \frac{1}{2}(AB + AC + BC)$.

Furthermore, we can also show that $PA + PB + PC < 2(AB + AC + BC)$.

To do this we need to use the triangle inequality repeatedly in figure 9-17 to show that $AB + AC > BP + PC$.

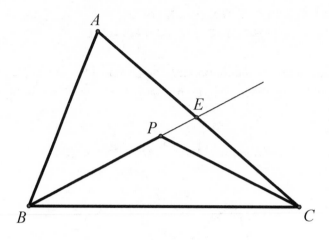

FIGURE 9-17

For triangle $PEC$, we have $PE + EC > PC$. Now follow along as we use the triangle inequality on various triangles in this configuration.

$$BP + PC < BP + PE + EC$$
$$BP + PC < BE + EC$$

However, $BE < AB + AE$. Therefore, $BP + PC < AB + AE + EC$, or $BP + PC < AB + AC$.

Applying this last relationship three times to triangle $ABC$ in figure 9-16, we find the following:

$$PA + PB < CA + CB$$
$$PB + PC < AB + AC$$
$$PC + PA < BC + BA$$

Adding these three inequalities we get

$$2(PA + PB + PC) < 2(AB + AC + BC),$$

which leads us to $PA + PB + PC < AB + AC + BC$.

Putting all of this together, we find that the perimeter of a triangle is less than twice the sum of the distances from any point inside a triangle to the vertices, yet greater than this sum. Symbolically, we would write this as $2(PA + PB + PC) > AB + BC + AC > PA + PB + PC$.

A further relationship involving the distances from a point, $P$, within a triangle to the vertices was discovered in 1902 by the Dutch mathematician J. N. Visschers, who showed that with the assumption that $AB$ is the shortest side of triangle $ABC$ (figure 9-18), we have $PA + PB + PC < AC + BC$. Only when $P$ coincides with a vertex will we have the following: $PA + PB + PC = PA + PB + CC = PA + PB = AC + BC$.

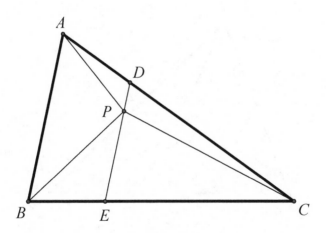

FIGURE 9-18

To prove this we begin by accepting that in figure 9-18, $AB \le BC \le AC$. We then draw a line through point $P$ and parallel to $AB$ that gives us $\triangle ABC$ ~ $\triangle DEC$. Whereupon it follows that $DE \le EC \le DC$ and $PC < CD$.

Applying the triangle inequality to the various triangles, we get $PA < AD + DP$ and $PB < PE + EB$ as well as $DE < CD + EC$.

From this we get

$PA + PB + PC + DE < (AD + DP) + (PE + EB) + (CD + EC)$
$= (AD + CD) + (EC + EB) + (DP + PE)$.

However, $PA < AD + DP$, $PB < PE + EB$, $PC < CD$, and $DE \le EC \ (\le DC)$.

Therefore, $PA + PB + PC + DE < AC + BC + DE$, so that by subtraction of $DE$, we then have $PA + PB + PC < AC + BC$, which is what we set out to show.

As we return to the randomly selected point in a triangle, we will consider the perpendicular distances to the three sides of the triangle. In figure 9-19, we get the following relationship: $PA + PB + PC \geq 2(PD + PE + PF)$, which was posed by Paul Erdös (1913–1996) and later solved by Louis Joel Mordell (1888–1972) and is known as the *Erdös-Mordell inequality*.[2] We should note that the equality here holds only when point $P$ is the centroid of an equilateral triangle. The proof is provided in the appendix.

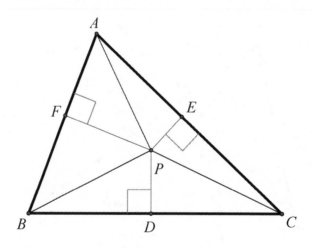

FIGURE 9-19

Another relationship attributed in 1962 to Louis Joel Mordell is one that can also be obtained from this configuration. It is as follows:

$$PA \cdot PB \cdot PC \geq (PD + PE)(PD + PF)(PE + PF).$$

## A CURIOSITY IN THE
## EQUILATERAL TRIANGLE

There is a curious inequality that evolves from a point inside (or even outside) a given equilateral triangle. Namely, that the three distances from this randomly selected point to the three vertices of the equilateral triangle will—with one exception—be such that they can determine a triangle.[3] That is, the sum of any two of these distances will be greater than the third. The exception is when the point lies on the circumscribed circle of the equilateral triangle. In that case, the sum of two distances will be equal to the third distance and no triangle can then be formed with these lengths.

In figure 9-20a, we have the following inequalities:

$$PA + PB > PC$$
$$PB + PC > PA$$
$$PA + PC > PB$$

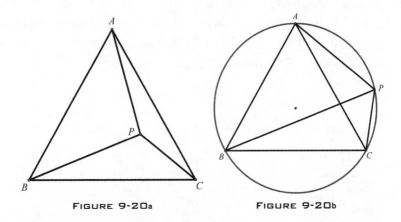

FIGURE 9-20a          FIGURE 9-20b

The astronomer Claudius Ptolemaeus, popularly known as Ptolemy (ca. 90 CE–ca. 168 CE), in his famous book, the *Almagest*, included a geometric theorem that he used to generate trigonometric values. Ptolemy's theorem states that for any four points $A$, $B$, $C$, and $P$, where no three of which are collinear, the following holds:

$$AB \cdot CP + BC \cdot AP \geq AC \cdot BP,$$

as may be seen in figure 9-20b. Since we are working with an equilateral triangle, $AB = BC = AC$. Therefore, $CP + AP \geq BP$. However, when $P$ is on the circumscribed circle of triangle $ABC$, we have a cyclic quadrilateral, and there Ptolemy's theorem states that $AB \cdot CP + BC \cdot AP = AC \cdot BP$.

Yet, when $AB = BC = AC$, as is the case for equilateral triangle $ABC$, we get $CP + AP = BP$. This is the one exception mentioned above.

## SOME INEQUALITIES
## IN THE RIGHT TRIANGLE

One inequality in a right triangle that can be practically "proved" by simply looking at it can be seen in figures 9-21a and 9-21b. Consider the right triangle $ABC$ with legs $a$ and $b$, and hypotenuse $c$. We can see that $a + b \leq c\sqrt{2}$, since the shortest distance between the two parallel sides of the square is the perpendicular distance, namely, $a + b$, which is clearly less than the hypotenuse, $c\sqrt{2}$, of the shaded right triangle in figure 9-21a. In figure 9-21b we see the case when $a + b = c\sqrt{2}$. Using the Pythagorean theorem on the shaded isosceles right triangle in figure 9-21a, we get $c = \sqrt{a^2 + b^2}$, which then gives us $\sqrt{a^2 + b^2} < a + b \leq \sqrt{2} \cdot \sqrt{a^2 + b^2}$.

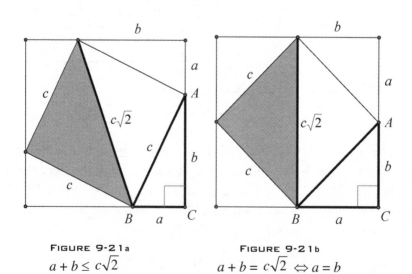

FIGURE 9-21a

$$a + b \leq c\sqrt{2}$$

FIGURE 9-21b

$$a + b = c\sqrt{2} \Leftrightarrow a = b$$

If we now introduce the radius of the inscribed circle of a right triangle, we have another unexpected inequality, namely, that the diameter of this circle is less than half the length of the hypotenuse of the right triangle. In figure 9-22, right triangle $ABC$ with legs $a$ and $b$ and hypotenuse $c$, we would have the following inequality, when $r$ is the radius of the inscribed circle: $2r \leq c\left(\sqrt{2}-1\right)$.

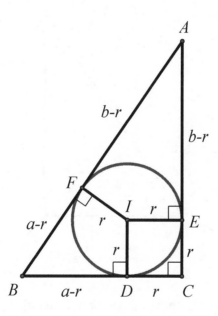

**FIGURE 9-22**

To justify this inequality is a relatively easy matter. With the points of tangency $D$, $E$, and $F$, we have $AF = AE = b - r$, $BD = BF = a - r$, and $CD = CE = r$. We then have the square $DCEI$ with side length $DC = CE = r$. It then follows that $c = AB = AF + BF = b - r + a - r = a + b - 2r$, or to put it another way, $2r = a + b - c$. We are now ready to use the earlier established inequality that $a + b \leq c\sqrt{2}$.

When we consider that $2r = a + b - c \leq c\sqrt{2} - c = c\left(\sqrt{2}-1\right)$, we have, in fact, demonstrated what we set out to do, namely, that $2r \leq c\left(\sqrt{2}-1\right)$.

There are a number of other inequalities that can be found involving a right triangle. One such is that the hypotenuse is never less than $\frac{1}{\sqrt{2}}$ times the sum of the legs. (See appendix.) In figure 9-23 this would be

$$c \geq \frac{1}{\sqrt{2}}(a+b) = \frac{\sqrt{2}}{2}(a+b).$$

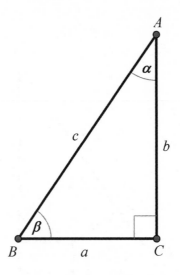

FIGURE 9-23

If we now draw the altitude $h_c$ to the hypotenuse of a right triangle as shown in figure 9-24, we get another surprising inequality, namely, that the hypotenuse is always less than $\frac{1}{2\sqrt{2}}$ times the sum of the legs of the right triangle. Symbolically that is

$$h_c \leq \frac{1}{2\sqrt{2}}(a+b) = \frac{\sqrt{2}}{4}(a+b).$$

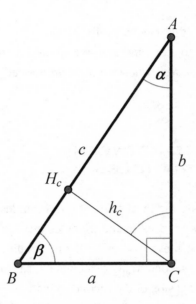

**FIGURE 9-24**

When we apply the previous inequality to each of the two smaller right triangles, we can state that

$$\Delta ACH_c: \quad b \geq \frac{1}{\sqrt{2}}(AH_c + h_c), \text{ and}$$

$$\Delta BCH_c: \quad a \geq \frac{1}{\sqrt{2}}(BH_c + h_c).$$

Adding the above two inequalities gives us

$$a + b \geq \frac{1}{\sqrt{2}}(BH_c + h_c) + \frac{1}{\sqrt{2}}(AH_c + h_c) = \frac{1}{\sqrt{2}}(AH_c + BH_c + 2h_c) = \frac{1}{\sqrt{2}}(c + 2h_c).$$

By simply replacing $c$ with $\frac{1}{\sqrt{2}}(a+b)$, we still can keep the inequality sign the same, since we already know that $\frac{1}{\sqrt{2}}(a+b) \leq c$, and we get

$$a + b \geq \frac{1}{\sqrt{2}}(c + 2h_c) \geq \frac{1}{\sqrt{2}}\left(\frac{1}{\sqrt{2}}(a+b) + 2h_c\right) = \frac{1}{2}(a+b) + \frac{2}{\sqrt{2}}h_c.$$

Now by solving this inequality for $h_c$ we get $h_c \leq \frac{1}{2\sqrt{2}}(a+b) = \frac{\sqrt{2}}{4}(a+b)$, which we set out to demonstrate. We should note that the equality embedded here holds true when we have an isosceles right triangle, that is, when $a = b$.

## INEQUALITIES BETWEEN A PAIR OF TRIANGLES

Here we will compare the third sides of two triangles when we are given that the remaining two sides of the triangles are respectively congruent. If the angle between these two sides of one triangle is greater than the included angle of the second triangle, then the third side of the first triangle is greater than the third side of the second triangle. In figure 9-25, we have $AB = DE$ and $BC = EF$. Then, if $\angle ABC > \angle DEF$, it follows that $AC > DF$. It should come as no surprise that the converse of this relationship is also true.

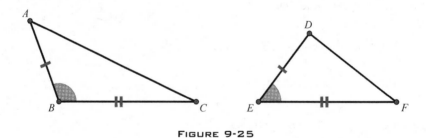

FIGURE 9-25

As an immediate consequence of this relationship we can show that in figure 9-26, where median $AM$ forms angles, such that $\angle AMC > \angle AMB$, it follows that $AC > AB$.

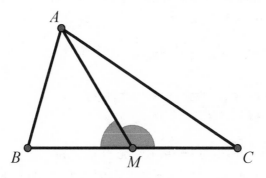

FIGURE 9-26

Since the sides forming the unequal angles (at point *M*) are respectively equal, the larger side is opposite the larger angle.

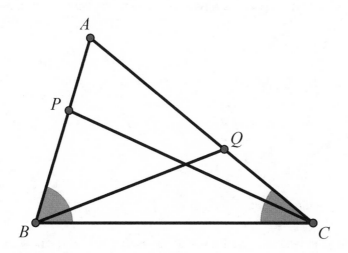

FIGURE 9-27

Sometimes the inequalities in a triangle can be a bit more difficult to justify. Here is one that will use a few of our previously encountered inequality relationships. In figure 9-27, we have triangle *ABC* with points *P* and *Q* on sides *AB* and *AC*, respectively. Also *BP = CQ* and *PC > QB*.

We must show that $AC > AB$. Applying the relationship established above to triangles $PBC$ and $QCB$, we get $\angle ABC > \angle ACB$. Then in triangle $ABC$, we have $AC > AB$.

## EVEN MORE INEQUALITIES IN A TRIANGLE

Inequalities in a triangle are to be found not only among the sides but also among the angles $\alpha$, $\beta$, and $\gamma$. Furthermore, there are interesting inequalities to be found in triangles when we also include the radii of the various circles related to triangles, such as the radii of the circumscribed circle $(R)$ , of the inscribed circle $(r)$, and of the escribed circles $(r_a, r_b, r_c)$. Adding the triangle's area gives even more such surprising inequalities.

### INEQUALITIES INVOLVING THE ANGLES OF A TRIANGLE:

$$\cos \alpha + \cos \beta + \cos \gamma \leq \frac{3}{2}$$

$$\sin \alpha + \sin \beta + \sin \gamma \leq \frac{3\sqrt{3}}{2}$$

$$\sin \frac{\alpha}{2} \cdot \sin \frac{\beta}{2} \cdot \sin \frac{\gamma}{2} = \frac{r}{4R} \leq \frac{1}{8}$$

$$\frac{a}{\sin \alpha} + \frac{b}{\sin \beta} + \frac{c}{\sin \gamma} \geq 12r$$

$$\sin^2 \alpha + \sin^2 \beta + \sin^2 \gamma \leq \frac{9}{4}$$

$$\tan^2 \frac{\alpha}{2} + \tan^2 \frac{\beta}{2} + \tan^2 \frac{\gamma}{2} \geq 1$$

### INEQUALITIES INVOLVING THE RADII OF A TRIANGLE:

The perimeter, $p$, of triangle $ABC$ with sides of length $a$, $b$, and $c$ is always less than six times the radius of the circumscribed circle, $R$. (See figure 9-28.)

$$a + b + c < 6R, \text{ or } \frac{a+b+c}{6} < R.$$

This inequality is very easy to justify.

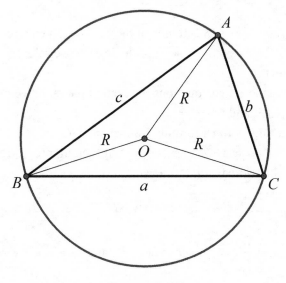

FIGURE 9-28

To justify this we focus on the three triangles $AOB$, $BOC$, and $COA$ and apply to each the triangle inequality, so we get the following:

$$2R = BO + CO > BC = a,$$
$$2R = CO + AO > AC = b,$$
$$2R = AO + BO > AB = c,$$

which by addition, gives us our desired result $6R > a + b + c$. This also holds true for right, acute, and obtuse triangles.

A famous inequality, relating the radii of the circumscribed and the inscribed circles, is often called *Euler's inequality* (1765), or *Chapple's inequality* (1746) or even sometimes the *Chapple-Euler inequality*.[4] It simply states that $R \geq 2r$.

The proof of this relationship can be rather simply done if we use area formulas that involve these radii. We presented these formulas in chapter 7 as follows:

$$Area\ \Delta ABC = \frac{abc}{4R}\quad (R = \text{radius of the circumscribed circle})$$

$$Area\ \Delta ABC = r \cdot s\quad (r = \text{radius of the inscribed circle, and}$$

$s = $ semiperimeter)

If we solve these two equations for each of the radii, we get

$$R = \frac{abc}{4\,Area\Delta ABC}, \text{ and } r = \frac{Area\Delta ABC}{s}.$$

Now recalling Heron's formula from chapter 7:

$$Area\ \Delta ABC = \sqrt{s(s-a)(s-b)(s-c)}, \text{ and making proper replacements}$$

and dividing the two equations, we have

$$\frac{r}{R} = \frac{\dfrac{Area_{ABC}}{s}}{\dfrac{abc}{4\,Area_{ABC}}} = \frac{4\cdot(Area_{ABC})^2}{abc\cdot s} = \frac{4\cdot s(s-a)(s-b)(s-c)}{abc\cdot s} = \frac{4\cdot(s-a)(s-b)(s-c)}{abc}.$$

Since $s - a = \frac{1}{2}(b+c-a)$, $s - b = \frac{1}{2}(c+a-b)$, and $s - c = \frac{1}{2}(a+b-c)$, it follows that

$$\frac{r}{R} = \frac{4\cdot(s-a)(s-b)(s-c)}{abc} = \frac{4\cdot\frac{1}{2}(b+c-a)\cdot\frac{1}{2}(c+a-b)\cdot\frac{1}{2}(a+b-c)}{abc} = \frac{(b+c-a)\cdot(c+a-b)\cdot(a+b-c)}{2abc}.$$

From our earlier discussion, we can make a replacement in the numerator as follows $(a + b - c)(b + c - a)(c + a - b) \leq a \cdot b \cdot c$, which gives us the following inequality:

$$\frac{r}{R} = \frac{(b+c-a)\cdot(c+a-b)\cdot(a+b-c)}{2abc} \leq \frac{abc}{2abc} = \frac{1}{2},$$

which then enables us to achieve the desired result: $R \geq 2r$.

We should note that the equality $R = 2r$ holds true when $a = b = c$, that is, when the triangle is equilateral.

We can sharpen up the *Chapple-Euler inequality* $\left(\frac{R}{2r} \geq 1\right)$ with the *Emmerich inequality* (1895), which states that $\frac{R}{2r} \geq \sqrt{2}+1$.

Also the following inequalities are universally valid:

$$\frac{R}{2r} \geq \frac{1}{3}\left(\frac{a}{b}+\frac{b}{c}-\frac{c}{a}\right) \text{ and}$$

$$\frac{R}{2r} \geq \frac{a^2+b^2+c^2}{ab+bc+ca}.$$

The famous German mathematician Gottfried Wilhelm Leibniz (1646–1716) discovered the following, which is therefore known as the *Leibniz's inequality*: $9R^2 \geq a^2 + b^2 + c^2$.

Furthermore, if the triangle is acute, then we get $9R^2 \geq a^2 + b^2 + c^2 \geq 8R^2$.

Here are some further inequalities involving parts of a triangle:

$$\frac{R}{2} \geq \frac{a+b+c}{6\sqrt{3}} \geq r$$

$$\frac{1}{ab} + \frac{1}{bc} + \frac{1}{ac} \geq \frac{1}{R^2}$$

$$a + b + c \leq 3\sqrt{3}\,R$$

Then there is also *Blundon's inequality*,[5] which has been popularized through mathematics challenges:

$$\frac{a+b+c}{2} \leq 2R + \left(3\sqrt{3}-4\right)r.$$

We can continue our list of triangle inequalities with the following:

$$\frac{1}{a} + \frac{1}{b} + \frac{1}{c} \leq \frac{\sqrt{3}}{2r}$$

$$|s^2 - 2R^2 - 10Rr + r^2| \leq 2(R - 2r)\sqrt{R(R-2r)}\,, \text{ where } s = \frac{a+b+c}{2}$$

$$24Rr - 12r^2 \leq a^2 + b^2 + c^2 \leq 8R^2 + 4r^2$$

If we now include the altitudes and the medians of a triangle we get the following surprising inequality:

$$h_a \leq \frac{b^2+c^2}{4R} \leq m_a, h_b \leq \frac{a^2+c^2}{4R} \leq m_b, \text{ and } h_c \leq \frac{a^2+b^2}{4R} \leq m_c.$$

And here is a simple inequality involving the altitudes and the radius of the inscribed circle:

$$9r \leq h_a + h_b + h_c, \text{ where the } \textit{equality} \text{ holds when } h_a + h_b + h_c = 9r.$$

Now involving the radius of the circumscribed circle we get the following inequality:

$$\frac{2}{\sqrt{3}} \le \frac{6R}{h_a + h_b + h_c}.$$

Yet we also have an inequality that involves only the altitudes of a triangle: $(h_a - h_b) \cdot h_c < h_a \cdot h_b < (h_a + h_b) \cdot h_c$.

The medians of a triangle give us the following inequalities:

$$m_a^2 + m_b^2 + m_c^2 \le \frac{27}{4}R^2$$

$$\frac{ab}{m_c^2} + \frac{bc}{m_a^2} + \frac{ac}{m_b^2} \le \frac{2R}{r}.$$

For the angle bisectors of a triangle, we have the following inequality:

$$\frac{t_a}{a} + \frac{t_b}{b} + \frac{t_c}{c} \ge \frac{a+b+c}{4r}.$$

Then we should also have an inequality that only involves the radii of the inscribed circle and the escribed circles of a triangle as follows:

$$\sqrt{r_a^2 + r_b^2 + r_c^2} \ge 6r.$$

Consider the following inequalities involving area.

For triangle $ABC$, where the side lengths are $a$, $b$, and $c$, and the area of the triangle is $Area \triangle ABC$, we get[6]

$$a^2 + b^2 + c^2 \ge 4\sqrt{3}\ Area \triangle ABC$$

and the even more impressive inequality[7]

$$a^2 + b^2 + c^2 \ge 4\sqrt{3}\ Area \triangle ABC + (a - b)^2 + (b - c)^2 + (c - a)^2.$$

Here are a few additional inequalities:

$$4\sqrt{3}\ Area \triangle ABC \le \frac{9abc}{a+b+c},\ \text{and}$$

$$\frac{9r}{2\,Area\Delta ABC} \le \frac{1}{a} + \frac{1}{b} + \frac{1}{c} \le \frac{9R}{4\,Area\Delta ABC}\,.$$

Prior to this chapter, most of our study of the secrets of triangles allowed us to relish the beauty of concurrency, collinearity, and above all, equality. Yet in this chapter we found that triangles expose some truly interesting and often-unexpected hidden gems in some consistent inequalities—ones that are true regardless of the shape or orientation of the triangle under consideration.

# CHAPTER 10

# TRIANGLES AND FRACTALS

W hat is a fractal and what can triangles tell us about them? These are two of the questions addressed in this chapter. We begin by introducing a very famous triangle responsible for shaking the very foundation of our beliefs about geometry.

## THE TRIANGLES OF PASCAL AND SIERPINSKI

In his *Traité du triangle arithmétique*, completed in 1654, the French mathematician and philosopher Blaise Pascal (1623–1662) described properties of a certain numerical, triangular array. Because of its numerous recursive relationships, many of which can be described in simple geometrical and arithmetical terms, this triangle has become a favorite for organizing the binomial coefficients that occur when a simple algebraic binomial is raised to a positive integral power. Consider

$(x + y)^0 = 1$
$(x + y)^1 = x + y$
$(x + y)^2 = x^2 + 2xy + y^2$
$(x + y)^3 = x^3 + 3x^2y + 3xy^2 + y^3$
and so on . . .

The coefficients of these algebraic expansions are sometimes seen arranged in an array as shown in the top panel of figure 10-1. In most Western countries, this array is referred to as *Pascal's triangle* in recognition of its inclusion in his paper. Since evidence of this triangle has been found in writings up to six centuries prior, and in widely separated regions of the world,[1] it seems certain that it and some of its properties were independently discovered several times over and likely had a variety of uses. Certainly, since modern algebraic notation would not have been in use in all those times, the construction of the array did not originally come from algebraic expansion, at least not in the form shown above.

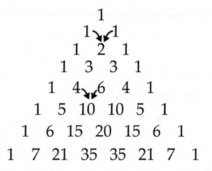

```
              1
            1 ⋁ 1
          1   2   1
        1   3   3   1
      1   4 ⋁ 6   4   1
    1   5  10  10  5   1
  1   6  15  20  15  6   1
1   7  21  35  35  21  7   1
```

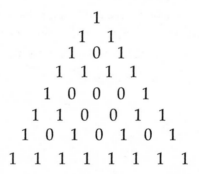

```
              1
            1   1
          1   0   1
        1   1   1   1
      1   0   0   0   1
    1   1   0   0   1   1
  1   0   1   0   1   0   1
1   1   1   1   1   1   1   1
```

FIGURE 10-1. EIGHT ROWS OF PASCAL'S TRIANGLE (*TOP*)
AND EIGHT ROWS OF THE 0-1 PASCAL TRIANGLE (*BOTTOM*).

Pascal's full array is not truly a triangle since it extends with infinitely many rows on one side. As a consequence, it will never be fully displayed.

When shown, it is given as a first few rows and perhaps some indication of infinite extent. If the tenth row of the triangle is known, then it is possible to write out the expansion of $(x + y)^9$ without resorting to the task of algebraic multiplication.

Other applications of Pascal's numbers abound. For example, if it is desired to determine the size of the task of listing all the five-card hands from a fifty-two-card deck, the answer depends upon the sixth array entry in row 53. Its value is the number of possible different hands, 2,598,960. Pascal's own use was in the discussion of various problems in probability.

All this would not be particularly useful without some computational method for quickly computing entries in the triangle. Fortunately, arithmetic relationships among the entries abound. Likely the simplest of these is that any non-boundary entry (the boundary numbers are all 1s) is the sum of the two entries just to the left and the right in the row above. The arrows in the top panel of figure 10-1 demonstrate two examples of this addition property. Armed with this fact and the left-right symmetry of the array it is possible to compute any entry in a fairly straightforward fashion.

In more recent years, stimulated in part by the needs of the digital age, attention has been drawn to a variant of Pascal's triangle that considers only whether the entries are odd or even (called the *parity* of an entry). This is most often presented by replacement of odd entries by the number 1 and even entries by the number 0, as seen in the bottom panel of figure 10-1. The summing relationship still applies, but with one adjustment. Since the sum of two even numbers is even, an entry is 0 if the two just above are 0 (e.g., use the fact that $0 + 0 = 0$). Since the sum of an odd number and an even number is odd, use $0 + 1 = 1 + 0 = 1$, as always. However, since the sum of two odd numbers is even, it is required to use a modified "addition" fact, $1 + 1 = 0$. Using these rules it is possible to quickly extend the 0-1 triangle downward to any depth.

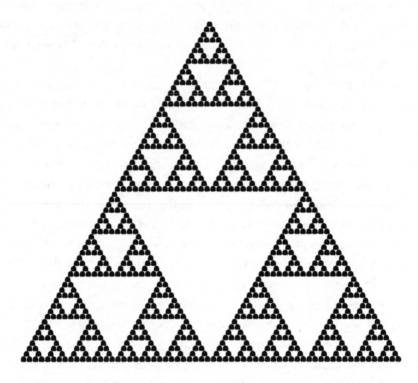

FIGURE 10-2. A PASCAL 0-1 TRIANGLE REDONE WITH DOTS AND
BLANKS, ALSO RECURSION LEVEL 5 OF TREMA REMOVAL.

A view of a large number of rows in the 0-1 triangle reveals an interesting geometric structure. Inverted triangles of zeros are seen and some organization of these by size is apparent. The effect can be portrayed even more clearly by replacing the 1s by black dots and the 0s by blank spaces. A Pascal triangle of sixty-four rows formatted in this manner is shown in figure 10-2. Interestingly this arrangement mirrors an almost-identical effect arising from a completely different area of mathematical investigation.

This same geometric pattern can be created without any reference to numerical data at all. The construction is the repetitive (or iterative) process pictured in figure 10-3 and next described.

0. Begin with any triangle (the *initiator*).
1. Join the midpoints of the three sides of the triangle to form a smaller, similar triangle in its interior and three other identical triangles at its corners.
2. Remove (or color in a contrasting color) the central triangle, leaving the three others.
3. Return to step 1, applying steps 1 and 2 to each of the remaining triangles, or exit the loop.

The steps just given describe a process that may be repeated for any number of passes through the loop (each pass is called an *iteration*, and the number of passes completed is called the *recursion level*). Recursion level 5 would create a copy of figure 10-2. Figure 10-3 illustrates the initiator and the outcomes from the first three iterations. The result from $n$ repetitions of this action mimics the look of $2^{n+1}$ rows of a 0-1 Pascal triangle.

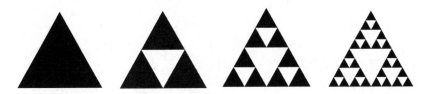

FIGURE 10-3. *LEFT TO RIGHT*, THE INITIATOR AND RESULTS AT RECURSION LEVELS 1, 2, AND 3 OF TREMA REMOVAL.

This triangle removal technique (*trema removal* from the Greek, meaning *hole* removal) was employed by the Polish mathematician Wacław Sierpinski (1882–1969) in 1915 when he introduced his famous triangle, the Sierpinski triangle. A Pascal triangle look-alike, it was devised to answer questions about the geometry of plane figures rather than to study the concepts investigated by Pascal. Sierpinski's geometric object is an example of a type of set called a *fractal*.

Just as no physical representation of the complete Pascal triangle can be displayed, a full Sierpinski triangle can be described only by an infinite generation method. The reasons vary for the two objects. The full Pascal triangle is not bounded on its bottom side so is not actually a triangle; it can never be completed in finite time by simply computing new rows. The full Sierpinski triangle is created only after infinitely many repetitions of triangle removals. In a sense, visualizing the Pascal triangle requires a telescopic view, looking outward ever farther. In contrast, the Sierpinski triangle is seen only by means of a microscopic view, working ever inward to greater degrees of magnification. These two mathematical entities exist only abstractly for the same reason that the positive integers can only be described but never fully exhibited. It was in the last part of the nineteenth century and the early part of the twentieth century that mathematicians made significant progress in understanding infinity and infinite processes. This was Sierpinski's era. The inscription on his grave maker reads, "investigator of infinity."

## THE MULTIPLE REDUCTION COPY MACHINE

Certainly, Sierpinski's triangle is interesting to construct, view, and contemplate. Since the construction process is very straightforward and simple, elementary-school children have successfully engaged in creating models of this object and clearly comprehend the basic method applied. But what have we in the end? Is the Sierpinski triangle "just a pretty face?" Some of the answer is that it is an extremely complicated structure, completely describable by a very simple repetitive process.

In the remainder of this chapter, we demonstrate a general process that constructs not only Sierpinski's triangle but also many other fractal (and non-fractal) objects. A variety of introductory examples will show its utility and some of the role of triangles in its use.

The following method is based on a process known, for introductory purposes, as a "multiple reduction copy machine" or MRCM. Not a physical machine, it is so named because its method can be described by analogy to a physical machine. We describe it by means of an example, beginning with a filled equilateral triangle, $\Delta$. By repeated application of the MRCM, we obtain a sequence of intermediate figures, $\Delta_1, \Delta_2, \Delta_3, \ldots,$

identical to those obtained by means of trema removal. The only difference will be in the process used to build them.

*Example 1.* We begin by making three copies of $\Delta$ while using a copy machine capable of reducing each original by a specified ratio. Also, we assume that the machine can do each reduction toward a specified point or center on the page, shrinking all the content toward this center. This center itself is not moved, but all other copied points move toward it. Finally, we assume that the original figure, $\Delta$, is printed on an overhead transparency and that the ensuing copies are also printed on similar material. While it will appear that only the printed content is affected, in reality, the machine examines every point of its scanning field. Transparent points result in transparent copies, so this part of the action of the machine goes temporarily unobserved. Reduction to half-size with the three vertices of $\Delta$ as the reduction centers is used to make the respective three copies. The original and the resulting copies are depicted in the panels of figure 10-4. The vertices of $\Delta$ are represented there by letters $A$, $B$, and $C$.

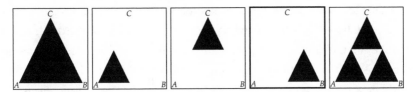

FIGURE 10-4. THREE REDUCED COPIES OF $\Delta$ COMBINE TO FORM $\Delta_1$.

The three copy sheets are superimposed and copied to create $\Delta_1$ on a single overhead transparency after the sheet containing $\Delta$ is removed. The final panel of figure 10-4 displays $\Delta_1$ and marks the completion of stage one of the MRCM's application. In a similar manner, using $\Delta_1$ as an original, $\Delta_2$ is created; then $\Delta_3$, and so on. As in the trema-removal method, this process only produces the Sierpinski triangle after an infinite number of repetitions. Our future reference to the MRCM will imply all those stages in the operation. Certainly a quite large copy-machine budget will be required!

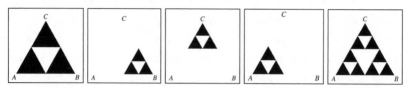

FIGURE 10-5. THE PROCESS OF CREATING $\Delta_2$ FROM $\Delta_1$ BY THE MRCM.

Even so, this method is likely to produce rather convincing results after a half-dozen, or so, repetitions. It is actually possible to carry out the first few stages of this experiment with a real copy machine, provided that it has a reduction feature and is given the ability to manually position the copies to simulate the three different reduction centers. A more satisfying result with less cost and more accuracy can be accomplished by use of computer software. A multitude of software applications exist that can create the Sierpinski triangle and other fractals. Many are free; some are expensive. Any list of these will omit someone's favorite.

Educators have used dynamic geometry software such as the Geometer's Sketchpad®, Cabri Geometry®, or GeoGebra® to construct fractals. Instructions and lesson plans for doing so are easily found in an Internet search. A software choice that allows mimicking the MRCM method would be a good companion to a reading of this chapter. Applications that support geometric transformations on the content of multiple hierarchical layers are especially suited for the task since layers can be used to implement the overhead-transparency idea. Adobe Photoshop® works beautifully for this purpose without a need for program writing, but other less expensive image-editing software can also be employed. The open-source GIMP® application is another of the many possibilities.

An MRCM produces equivalent results to trema removal in some cases, the example just displayed being one of those. However, we will demonstrate examples in the following that show its potential to far exceed the removal method's capabilities.

## ITERATED FUNCTION SYSTEMS
## AND FIXED SETS

Once the operational procedure and the reduction centers are established, an MRCM can act on any image, not just the triangle we first proposed. In order to describe the multitude of possibilities, the copy-machine analogy is generalized in the following way. We will consider an MRCM to be a process based upon a list of one or more geometric transformations. A different list of transformations establishes a different MRCM. The list of transformations and their use in an MRCM is also commonly called an *iterated function system* or IFS.

Each transformation acts to convert an image (a set of points) into a second image by following some geometric rule. In the early examples, these transformations were *contractions* that shrink an image by some specified ratio toward a central point. Other transformations can include rotation of an image about a point or reflection about a line. Any collection of these actions can also be combined into a single, even more complex transformation.

So that they can be discussed individually, the transformations in an IFS will be denoted by a subscripted character, $F$, so that the list of an IFS will look like $F_1, F_2, F_3, \ldots, F_k$, where $k$ is the number of transformations in the IFS. A single iteration of the MRCM (or IFS) then consists of the separate application of each of the transformations, $F_i$, to the current version of the image and the consolidation of these separate transformed images into a single combined image by overlaying the transformed parts. The effect of full implementation of the machine is to produce an infinite sequence of images, emanating from an initial beginning image called an *initiator*. The initiator is not considered a defining part of the MRCM but simply one of the many sets upon which it may act.

Example 1 can be rephrased in this language. The IFS in example 1 is based on three transformations, $F_A$, $F_B$, and $F_C$. Each of these transformations shrinks any image by a ratio $\frac{1}{2}$ toward its respective center $A$, $B$, or $C$. The first iteration of the resulting MRCM applied each of these contractions to the initiator, a filled triangle, $\Delta$, resulting in the three images shown in the middle three panels of figure 10-4 and then the combining of these by superposition to obtain $\Delta_1$ (called the *first iterate*). The second iteration applied $F_A$, $F_B$, and $F_C$ to $\Delta_1$ and combined those results to obtain $\Delta_2$. Third, fourth, fifth, and so on, iterates are obtained by repeatedly applying additional iterations. When completely applied, the MRCM will have generated an infinite sequence of iterates, $\Delta$, $\Delta_1$, $\Delta_2$, $\Delta_3$, . . .

Since there can be any number of transformations defining an IFS and many different initiators, sequences of iterates can be quite varied. The next several examples show some of the possibilities and are intended to promote a deeper understanding of IFS properties.

*Example 2*. What happens if the same IFS as used in example 1 is applied to some different initiator set? Consider O to be a circle (and its interior) covering the triangle, $\Delta$, of example 1. Our interest is then in seeing how $O$, $O_1$, $O_2$, $O_3$, . . . compare to our first result. The figure 10-6 tells this story. In panel 2 we see the merged result of the three transformed versions of the circle, O, completing iteration step 1. Iteration 2 is not pictured, but the three corner triangles in panel 3 are the reduced versions of $O_2$ that combine to form $O_3$.

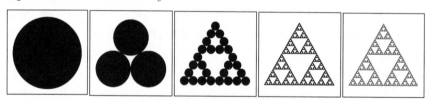

FIGURE 10-6. LEFT TO RIGHT, O, $O_1$, $O_3$, $O_6$, AND $O_8$ SHOWING CONVERGENCE TO THE SIERPINSKI TRIANGLE.

The result, taking shape in the final panels, is far different than might have been expected! It would be natural to expect that the triangular shape with which we began our example 1 was primarily responsible for the triangular shape of the Sierpinski triangle. Following that line of reasoning it might be expected that example 2 would produce a Sierpinski circle. What we actually obtained motivates looking at examples from other initiators.

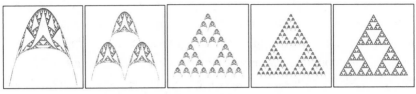

FIGURE 10-7. *LEFT TO RIGHT*, $\Lambda$, $\Lambda_1$, $\Lambda_3$, $\Lambda_4$, AND $\Lambda_9$
AGAIN SHOWING CONVERGENCE TO THE SIERPINSKI TRIANGLE.

*Example 3.* Figure 10-7 shows a quite different initiator and the results at several iteration stages. By the ninth iteration, we have an image with very much the appearance of the Sierpinski triangle. The Sierpinski triangle is persistent. It turns out that nearly *any* non-empty initiator will lead to it when this particular IFS is applied. Amazingly, even a starting set consisting of only a single point will yield the same final outcome.

Some initiators do not lead to this same result. For example, an ordinary horizontal line through the center of the original triangle will result in a horizontal strip of a thickness equal to the height of the triangle. It is known that all initiators that are *bounded* (e.g., can be contained within a circle of finite radius) will always lead to same single result for a given (*contractive* as described below) IFS. The line just described is not a bounded set.

If the shape resulting after infinitely many iterations is not determined by the nature of the initiator used, then what is causing the Sierpinski triangle to turn up each time? One does not have to look far for the answer. The only part of our process remaining to suspect is the selection of the IFS itself. It turns out that its transformations contain a blueprint for the Sierpinski triangle just as the genetic code of an animal encrypts the essential form of that creature. The transformations $F_A$, $F_B$, and $F_C$ are the "DNA" of our IFS. Any different selection of transformations can be expected to create a corresponding change in the *limit set* (i.e., the ultimate resulting image after infinitely many iterations of an IFS). That is not to claim that all transformation selections can be expected to settle to a stable limit.

Consider some alternate choices for the IFS rather than a different initiator. What if it is defined with fewer transformations? The transformations chosen can be quite general but for now we restrict ourselves to *contractive* transformations. Roughly speaking, such transformations have the property that image points move closer together each time the transformation is applied. Shrinking an image toward a fixed point and to one-half its original size is an example of a contractive transformation. Under this restriction, it is known (Michael Barnsley, *Fractals Everywhere*, 2nd ed. [Boston, MA: Academic Press, 1993], pp. 74–81) that a stable limit set will always result when the IFS is applied to a permissible initiator.

*Example 4*. Suppose we use only the two transformations $F_A$ and $F_B$ of example 1. Figure 10-8 shows how things proceed with a filled circle, O, as an initiator. The limit set is, as it appears in the last panel, a straight line segment. The segment's endpoints are the points A and B. Choosing to use only any two of the three transformations from example 1 produces a line segment. It seems that all three transformations are necessary in order to generate the Sierpinski triangle.

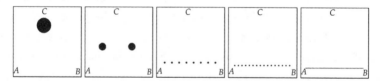

FIGURE 10-8. *LEFT TO RIGHT*, O, $O_1$, $O_3$, $O_4$, AND $O_{11}$
SHOWING A LINE SEGMENT AS THE LIMIT SET.

*Example 5*. What happens if only one transformation, $F_A$, is used? The portion of each iteration that calls for combining the transformed images then becomes essentially trivial. For an initiator, T, the result, $T_1$, is simply the same as the result of applying $F_A$ to T. Figure 10-9 shows the result. We will find that, after infinitely many iterations, the limit set is a single point, the point A. Even without conducting an experiment, it is relatively easy to see why this would be the case.

Since the action of $F_A$ is to move points of any image half the distance toward the point A, applying $F_A$ to the point A would again produce A. When this happens, we call the point a *fixed point* of the transformation, so the special occurrence in this example is that the limit set is also a fixed point of the MRCM. This is an exceptional situation. If an MRCM has a limit set that is not a single point, then an application of a single iteration of the MRCM to that set will move its points about, but only within the limit set. Usually this effect cannot be seen in a computer-generated image since the limit *set* remains unchanged. In general, when an MRCM has a limit set, that set is a *fixed set* of the MRCM but only in exceptional cases a fixed point.

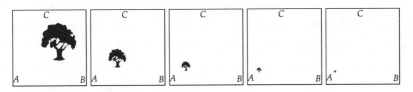

FIGURE 1 0-9. *LEFT TO RIGHT*, T,T$_1$,T$_2$,T$_3$, AND T$_4$
SHOWING A SINGLE POINT AS A LIMIT SET.

Here is a summary the general situation. For any chosen positive integer, $k$, a list of transformations, $F, F_1, F_2, F_3, F_4,..., F_k$, determines an IFS that can be iterated through use of an MRCM method. The MRCM can be applied to any initiator image, $\Delta$, that is a *closed* and *bounded* set of points (i.e., edges are included and the set is not of infinite extent) to produce an infinite sequence of sets, $\Delta, \Delta_1, \Delta_2, \Delta_3, \ldots$, called iterates. When the transformations of the IFS are all *contractive* then the iterates approach ever closer to a limit set that is a fixed set of the MRCM.

In our examples thus far we have seen $k = 1, 2$, and 3. The initiators used were a filled equilateral triangle, a filled circle, a drawing of a tree, and one other less familiar shape. The transformations, $F_i$, used were all contractive in that for every pair of distinct points, $P$ and $Q$, the distance between their transformed images did not exceed a fraction of the distance between $P$ and $Q$. Since, in our examples, this ratio was exactly one-half, we say that one-half is the contraction factor of each $F_i$. The limit sets were, respectively, the Sierpinski triangle, a line segment, and a point.

The results up to now are beginning to appear very orderly. Three transformations produced a triangle, two produced a line segment, and one produced a single point as the fixed set. Actuality is not so simple. The next sequence of examples shows that the number of transformations is not, of itself, a predictor of the shape of a limit set.

*Example 6.* For $k = 2$ (i.e., two transformations) if we take both $F_1$ and $F_2$ to be the same as $F_A$ from example 5 and $T$ the initiator from that example, then the iterates will be the same as in example 5. The point $A$ will be the limit set and the diagram of figure 10-9 still applies.

If $k = 3$ and the same three transformations are used as before except that we choose three reduction centers elsewhere on the plane rather than as vertices of an equilateral triangle, then the limit sets derived will be easily identified triangles that look like modified, transformed versions of the Sierpinski triangle. Standardly, the title, *Sierpinski triangle*, is taken to mean any one of these. It can be, however, that the three reduction centers are collinear. In such a situation, the limit set will be a line segment (or possibly even a single point).

*Example 7.* A more interesting example can be shown for $k = 4$. This example takes $F_A$, $F_B$, and $F_C$ just as before with fixed points at the vertices of a triangle, $\Delta$. The description of the additional transformation, $F_G$, is slightly more complex. The three *medians* of a triangle are the lines joining the vertices of the triangle with the midpoints of the opposite sides. It has long been known that these lines are concurrent at a single point, $G$, called the *centroid* of the triangle (see chapter 2). The centroid, for many purposes, serves as a center point of the triangle. The action of $F_G$ will be to first rotate the initiator by 180° about $G$ and *then* follow with a contraction by the ratio one-half toward $G$. The effect of $F_G$ is seen in panel 4 of figure 10-10. The entire figure shows what happens when the first iteration of the MRCM is applied to a filled triangle, $\Delta$. Dashed boundary lines have been added to expose the relative positions (in relation to $\Delta$) of the four parts that combine to form $\Delta_1$.

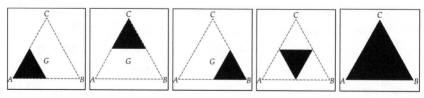

FIGURE 10-10. *LEFT TO RIGHT*, THE ACTION ON $\Delta$ OF $F_A$, $F_B$, $F_C$, AND $F_G$ AND THEIR COMBINATION, $\Delta_1$.

This figure demonstrates only one iteration of the MRCM, but it tells the whole story. We see that $\Delta_1 = \Delta$, so $\Delta_2 = \Delta$, and so on, and hence that $\Delta$ is a fixed set and limit set of the MRCM. Perhaps this is a little disappointing. No new and interesting limit set was obtained. It is a little more interesting that a circle or other initiator under this same IFS would also converge to $\Delta$. With $k = 4$, we might have hoped for something more exotic or, at the very least, some form with four vertices. The real secret of this example is that the four subsets in the first four panels of the figure are completely similar to $\Delta$ and together form a tiling of $\Delta$. Any tiling of a triangle by reduced copies of itself, using two or more tiles, can be used to create an example of this kind. A very large number of transformations in an IFS can, in this way, produce a simple triangle.

Perhaps the value of $k$ establishes an upper bound of a sort. Is it the case that $k$ must be 3 or more in order for an IFS to produce a triangle as its limit set? The answer is *no*, as is shown by the next example.

*Example 8.* An IFS composed from only two transformations can still produce a triangle. Imagine two points, $A$ and $B$, arranged in a horizontal line with $A$ on the left. The transformation, $F_A$, is taken as a reduction toward $A$ with a ratio of $\frac{1}{\sqrt{2}}$. Prior to the reduction, $F_A$ reflects an image about the line $AB$, and after the reduction it rotates an image about the center $A$ by 45° counterclockwise. The other transformation, $F_B$, acts in exactly the same manner, but with center at $B$ and a *clockwise* rotation by 45°. The limit set of the IFS determined by these transformations can be completely understood by considering the action of one iteration of the MRCM acting upon an isosceles right triangle situated so that the segment $AB$ is its base. Figure 10-11 shows the result of this action. Since the triangle is seen to be a fixed set under one iteration, it remains unchanged under all the following iterations and so the triangle is the limit set of the IFS.

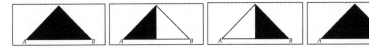

FIGURE 10-11. *LEFT TO RIGHT, T,* $F_A$ *APPLIED TO T,* $F_B$ *APPLIED TO T,* AND THEIR MERGER TO FORM $T_1 = T$.

From examples 4 through 8 we see that, although the value of $k$ contributes to the shape of an IFS limit set, the value of $k$ alone is not fully predictive of the shape. The examples 7 and 8 also introduced the idea that the transformations, $F_i$, need not be simple reductions toward a point but can involve other actions such as rotation and reflection. Such more complicated effects in three transformations can produce interesting "relatives" of the Sierpinski triangle. The next two examples show how this is done.

*Example 9.* Points $A$, $B$, and $C$ are assigned as the respective, top-left, bottom-left, and bottom-right corners of a square. If each of the three corresponding transformations, $F_A$, $F_B$, and $F_C$, with these points as reduction centers, simply have $\frac{1}{2}$ as a reduction ratio, then, as in example 1, the limit set will be a Sierpinski triangle, but now a right-triangle version. Following an idea from example 7, $F_A$ is modified by including, in its effect, a rotation by 180° about the center of the square *prior* to the 50 percent reduction toward $A$. The other two transformations are left as simple reductions. For this demonstration, we use a filled version of the right triangle, $T$, as an initiator. Once the three transformations are well understood, the appearance of $T_1$ is fairly easy to predict. The panels in figure 10-12 show several iterations and approximate the limit set.

FIGURE 1 0 - 1 2. *LEFT TO RIGHT,* T, T$_1$, T$_2$, T$_3$, AND T$_8$.

*Example 10.* Repetition of the last experiment, but with one more alteration, produces a different limit set. For this example, $F_C$, is augmented with an additional step, reflection of the square region about its vertical axis, then followed by the reduction toward $C$. Iterations are shown in figure 10-13.

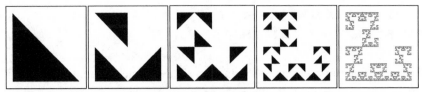

FIGURE 10-13. *LEFT TO RIGHT*, T, T$_1$, T$_2$, T$_3$, AND T$_9$.

It should now be apparent that a large number of different limit sets might be obtained in a similar manner. If one permits only reductions by 50 percent toward the three vertices so far identified and allows one symmetry motion of the square's contents to be applied in each transformation prior to the reduction, then $8^3 = 512$ different versions of the MRCM can be obtained. The number 8 arises since there are exactly eight ways to rigidly transform a square onto itself. Rotations in the amounts of 0°, 90°, 180°, and 270° are four of these. The other four are reflections about the square's four axes of symmetry. The total number of actually distinct outcomes is less than 512, since some different versions of the MRCM do not produce distinct limit sets. For example, if $F_B$ is constructed to first reflect the square about the diagonal that runs from lower left to upper right before reduction and if the other two transformations are simply the original reductions, then the limit set is exactly the Sierpinski right triangle as was obtained when $F_B$ was a simple reduction only.

In order to decide if two image sets among these cases are actually different it is necessary to first agree on the meaning of *different*. For example, if two limit sets are mirror images of each other, should we count them as distinct? One way of specifying "different" for these sets is to require that the limit sets be considered the same if any one of the eight symmetries of the square, applied to the containing square, can transform one into the other. Under this definition there are 232 possible distinct image sets resulting from the class of transformations specified when $k = 3$.

This discussion began by using a copy-machine metaphor that served to maintain a simple description for the transformations we have used. Now, as our examples have become somewhat more wide-ranging, the specification of the transformations has become more cumbersome. This is a good time to introduce a more efficient way to describe them.

## USING TRIANGLES TO
## DEFINE TRANSFORMATIONS

The selection of possible transformations in the plane for creating an IFS is hugely broad. Fortunately, there are specialized classes of transformations that are a fertile source of interesting examples. Those used in our examples to this point fall into two classes.

1.  *Isometries* or *Euclidean congruences* are the transformations that preserve distance and angle measure. That is, the distance between any two points is the same as the distance between their transformed images. *Rotations* about a point, mirror *reflections* about a line, and *translations* (directional shifts) all are in this category. The effect of any isometry upon an image can be successfully visualized as the result of shifting, rotating, or flipping an overhead transparency upon which the image is drawn.

2.  *Similarities* preserve the ratios between distances and angle measure. If the distance between points $a$ and $b$ is $k$ times the distance between $c$ and $d$, then the two distances between their corresponding, transformed image points bear this same relationship. The reductions toward a point in our preceding examples are similarities but the isometries are also similarities with a ratio of 1 (i.e., congruent figures are also similar figures).

When the transformations in the above two classes are applied to any triangle, the transformed result is also a triangle. When the transformation is an isometry, the result is a congruent triangle. When the transformation is a similarity, the result is a similar triangle. An even more useful, but less known, fact is that the complete nature of one of these transformations can be discerned from the knowledge of a single triangle and its transformed image. It is only necessary to know how the vertices of the original are matched to the vertices of the image. In essence, this works because the three vertices of a triangle contain sufficient information to establish a coordinate system for the plane. Armed with that information it is then possible to determine how the transformation affects any other point of the plane.

The matching between an original triangle and its image is often specified by making a list showing corresponding vertices, but in most cases, it is simpler to show a diagram of the two triangles marked with arrows that specify the matching. We adopt the convention of placing the arrow on the original image outside the figure and the arrows on the transformed images inside those triangles. Figure 10-14 shows how such diagrams would look in our previous examples 1 and 8. The arrow on a large triangle and an arrow on a smaller triangle show two vertices that must match for the transformation that transforms the bigger to the smaller (points at the beginning of the arrows match and points at the ends of the arrows match). Then the third (remaining) vertices must also match.

It is seen in the IFS of the second panel of figure 10-14, for example, that there are two transformations and that each must involve flipping the original triangle over (since the small arrows travel clockwise around their triangles in contrast to the large arrow which is directed counterclockwise). Shrinking and rotation is also required in each case. In the first panel, there are three transformations. No rotation or flipping is required, but each small triangle has sides that are half the length of the big triangle. Thus, each transformation in that case is a similarity with ratio $\frac{1}{2}$.

Representing transformations with oriented triangles works well in many cases but may be less satisfactory when very accurate representations of the transformations are needed. For those latter cases, specifying coordinates of the vertices and algebraic methods are usually used. Those not wanting to go through such an algebraic exercise will be comforted by the additional known fact that the similarities can always be described as a combination of rotations, reflections about lines, translations, and a single *dilation* (a rescaling about a fixed point). It often suffices to merely find such a combination. In most of the remaining examples, we will simply indicate arrows with triangles and their images to describe the transformations making up an IFS and its corresponding MRCM.

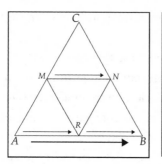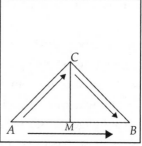

FIGURE 10-14

Figure 10-15 demonstrates how simpler geometric motions might be determined in the case of a similarity. It is possible that the axis of reflection, angle and center of rotation, direction of translation, and center of reduction could be selected to be different from those indicated and yet their combination could produce the same combined effect as the ones shown. Component actions might also be applied in a different order. For example, a rotation could come first. An extremely important point is that once the component parts are specified, the order of their application is crucial to the correct outcome. A different order with the same motions quite often produces a different composite transformation. It is also known that a single rescaling and a few reflections can always achieve a given similarity. No more than three reflections are ever required.

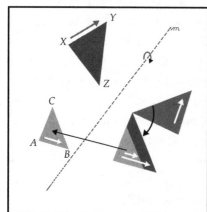

1. *Reflect around axis m*

2. *Rotate to make sides parallel*

3. *Contract toward a point to match size*

4. *Translate to the correct position*

FIGURE 10-15. MOTIONS TO TRANSFORM TRIANGLE XYZ TO A SIMILAR TRIANGLE ABC.

It is worth emphasizing that these transformations act not just upon the points inside and on the initial triangle but also extend their action to every point of the plane. It is this generality that permits applying the successive iterations of the MRCM for any initiator chosen.

One thing that becomes clear as a result of how triangle matching specifies a transformation and how these, in turn, specify an IFS is that any subdivision of a triangle into a collection of sub-triangles similar to the original can be used to create an IFS with the original triangle as its initiator. If all the sub-triangles are used in the definition, the original triangle will become the fixed set. Examples 7 and 8 are illustrations of this fact. Extensions of this idea can be made to non-triangular figures that can be tiled in a self-similar way. Rectangles, hexagons, pentagons, and even line segments are some of these figures. The Sierpinski triangle gives an example of how a very different proper subset of the initiator can turn out to be the fixed set.

The impact of all this is quite amazing. We now see that there is a large class of geometric objects, each of which can be completely described by a single IFS. In numerous important cases, the transformations making up the IFS can be completely specified by designating correspondences between the members of pairs of similar triangles. Research by Michael Barnsley and others has focused on the potential to use this fact for modeling complex natural objects in an efficient manner.

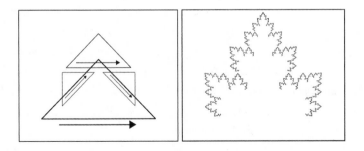

FIGURE 10-16. THE DEFINITION OF AN IFS WITH $\kappa = 3$ IS SHOWN IN THE LEFT PANEL AND ITS FIXED SET IN THE RIGHT PANEL.

## FRACTALS AND FRACTAL DIMENSION

Benoît Mandelbrot (1924–2010) introduced the term *fractal* in 1975 (Benoît Mandelbrot, *Fractals: Form, Chance and Dimension* [San Francisco, CA: W. H. Freeman, 1977]). Since then, thanks in part to the information-technology revolution, the word has entered mainstream vocabulary so as to be applied rather imprecisely to objects that exhibit rough approximations of characteristic fractal properties. The term is used especially loosely when applied to natural objects such as trees and plants. There is not full agreement, even in scientific circles, about the formal definition of *fractal*. Rather than attempt to settle the issue in this chapter, we simply focus on two fractal-related properties, self-similarity and fractal dimension.

Self-similarity in fractals has been described as "repetition across scale." A rather simple form of this occurs in optical feedback, the effect seen when a video camera is directed at a video monitor connected to the camera's output. The resulting video image, sometimes called the *Droste effect*, shows a recursive picture-within-a-picture receding to apparent infinite extent. Any enlarged and cropped version of the display does not appear different from the original. The Sierpinski triangle has exactly this property but in a more complex form. If the abstract Sierpinski triangle is cropped to include only one of its three corner sub-triangles and the result is then enlarged to double its scale, an exact copy of the Sierpinski triangle is obtained. Fractal images produced using an IFS will always have self-similarity due to the nature of the MCRM iteration process.

While some would simply adopt self-similarity as the defining property of fractal, this introduces some unwelcome consequences. Some objects that we might wish to have be fractals only have a self-similarity property in a very approximate form. On the other hand, some objects that would not commonly be considered as fractal do have precise self-similarity. Kazimir Malevich's[2] painting *The Black Square* would be a painting of a fractal under this view. Some of the examples above created filled triangles as self-similar sets.

There is common understanding of what is meant by a one-, two-, or three-dimensional object but to propose a fractional dimension such as 1.3 will, for many, appear as nonsense. Even so, proposing a definition of geometric dimension that accommodates fractional values is exactly what mathematicians have done. How dimension is treated has almost everything to do with how length, area, and volume are measured.

There is a long history in the struggle to understand the area concept in geometry. Geometric measurements are mostly not absolute but rather comparative. One line segment can be described as congruent to another and hence of the same length. Thus, length is measured by comparing with a standard unit segment.

Dissection is also important. When a segment can be dissected into two congruent segments of length 1, the original is said to have length 2. This idea is the basis of how we apply linear measure.

Geometric transformations bear a fundamental relationship to the application of measurement. Isometries do not change distance when they are applied and hence by Euclid's SSS congruence theorem they also do not change triangle area. Dependence on triangles is fundamental to area measurement of all polygons since they can be dissected into triangles. Isometries thus preserve areas of transformed objects as well as lengths.

Transformations that contract toward a central point maintain angle size and the ratio of corresponding lengths. This is fundamental to the elementary geometry notion of similarity. Combined with the area-preservation property of isometries, this fact implies that similarities reduce (or increase) area proportionately across different geometric shapes.

When a square is dissected into four congruent sub-squares with sides half the original square's side length, their areas must be $\frac{1}{4}$ the original area. If, instead, the side length division is to $\frac{1}{3}$ the original, then there are nine sub-squares of area $\frac{1}{9}$ the original (A. S. Posamentier and I. Lehmann, *The Glorious Golden Ratio* [Amherst, NY: Prometheus Books, 2012], pp. 269–92).The general relationship is that $k^2 = n$ when $\frac{1}{k}$ is the reduction ratio and $n$ is the number of congruent pieces. In general, this relationship extends to areas of all geometric shapes.

This feature has generally been taken as characteristic of two-dimensional objects. When the object is contracted by a similarity ratio of

$\frac{1}{k}$, then its area is contracted by $\frac{1}{k^2}$. For one-dimensional objects the ratio is $\frac{1}{k}$, and for three-dimensional objects it is $\frac{1}{k^3}$. The relationship covering all of these is

$$\left(\frac{1}{k}\right)^d = \frac{1}{n},$$

where $\frac{1}{k}$ is the similarity ratio, $n$ is the number of congruent similar parts in a dissection, and $d$ is the dimension.

This understanding makes it possible to see what is truly amazing about the Sierpinski triangle. It was constructed using a contraction ratio of $\frac{1}{2}$, but the resulting object is correspondingly dissected into three exact copies of itself. The relationship here is

$$\left(\frac{1}{2}\right)^{1.53...} = \frac{1}{3},$$

distinguishing the Sierpinski triangle from both one- and two-dimensional objects.

The value 1.53 . . . in the above was calculated by solving the equation

$$\left(\frac{1}{k}\right)^d = \frac{1}{n}$$

for $d$ by using logarithms. This leads to the formula,

$$d = \frac{\log n}{\log k},$$

for the dimension of self-similar objects when all the similar sub-parts are congruent. When this is not true, the self-similarity dimension can still be calculated but with slightly less ease.

Revisiting the dimension calculation for a filled square supports the conclusion of this formula. The square can be regarded as self-similar by considering it as the union of the four sub-squares obtained by bisecting the square in both directions. Thus, the value of $n$ is 4 and the reduction ratio is $\frac{1}{b} = \frac{1}{2}$. The dimension calculation then yields

$$d = \frac{\log 4}{\log 2} = \frac{\log 2^2}{\log 2} = \frac{2\log 2}{\log 2} = 2 ,$$

exactly the traditional dimension for a filled square.

A line segment can be regarded as the union of two line segments, each half its length. We thus have $n = 2$ and $\frac{1}{b} = \frac{1}{2}$ so

$$d = \frac{\log 2}{\log 2} = 1 ,$$

again what we expect.

Nonintegral dimension is typical of the geometric objects called *fractals*. Indeed some would go so far as to propose fractional dimension as the defining property of fractals.

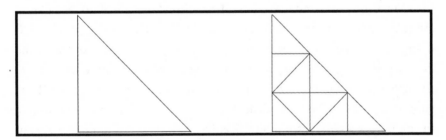

FIGURE 10-17

*Example 11.* An isosceles right triangle, $\Delta$, can be tiled in a self-similar fashion by nine smaller isosceles right triangles, as shown in figure 10-17. Any number from one to nine of these smaller triangles may be selected and used with the large triangle to define transformations, creating an IFS. Each of these transformations would have a similarity ratio of $\frac{1}{3}$. Each such IFS has a self-similar fixed set and we can compute the dimension of this geometric object. In general, the dimension is

$$d = \frac{\log k}{\log 3},$$

where $k$ is the number of transformations, so the list of possible dimensions for a fixed set obtained in this way is

0, 0.631, 1.0, 1.262, 1.465, 1.631, 1.771, 1.893, and 2.0.

The effect of one of the resulting operators applied to a filled version of $\Delta$ and with $k = 6$ is shown in figure 10-18. By the fifth iteration of the MRCM the structure of the resulting fractal has become rather clear. The six small triangles to which $\Delta$ is mapped define transformations that can be observed in the second panel of the figure 10-18. The dimension in this case is 1.631, slightly larger than the 1.585 of the Sierpinski triangle. A possible inference from this result is that the fractal produced is slightly closer to being a two-dimensional object than is the Sierpinski triangle.

One of surprises hidden among the multitude of possible examples is the case for $k = 3$ where the dimension is 1.0. Some choices of three small triangles produce results as expected for dimension 1. For instance, if the three small triangles whose hypotenuses lie on the hypotenuse of the large triangle are chosen and $k = 3$, then the limit set is the hypotenuse of the large triangle. For this line segment, the dimension 1 seems appropriate. In contrast, when the three triangles at the vertices of the large triangle are selected the result appears to be somewhat fractal-like and the dimension 1 is more surprising.

FIGURE 10-18. *LEFT TO RIGHT*, $\Delta$, $\Delta_1$, $\Delta_2$, $\Delta_3$, AND $\Delta_5$.

Other difficulties with self-similarity dimension must be dealt with. Not all the sets that we want to describe as fractals have a simple detectable self-similar generation method. For these an alternate definition of *dimension* may be used. Conflicting results have been observed for some example images under the various versions of dimension. Another trouble spot is that two distinct IFSs can sometimes generate the same limit set, leading to a potential disagreement of values.

## ORBITS AND ORBITAL SETS

There are more aspects of interest about an IFS than its fractal attractor. The collection of all *iterates* (the sets obtained at each stage of iteration) is called the *orbit* of the IFS. The union of the sets in the orbit of the IFS is called the *orbital set*. A simple example of such an orbital set is seen in the right panel of figure 10-20. The IFS is composed of a single transformation as defined in the left panel and the limit set is simply the single fixed point of that transformation that lies at the apex of the initial triangle. The orbital set in this case is more interesting than the limit set. The collection of all the individual flowers comprises the orbit of the IFS. In general, pictures of orbital sets are a useful means for communicating a summary of how the iterates of an IFS change across iteration levels. Many are also quite beautiful and are potential content for works of art. Many such examples exist in the works of M. C. Escher (1898–1972), a Dutch graphic artist who is known for his "impossible dipictions."

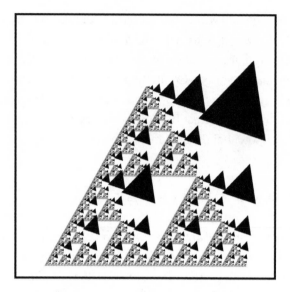

FIGURE 10-19. AN ORBITAL SET DEMONSTRATING A SMALL TRIANGLE
TRYING TO BECOME THE SIERPINSKI TRIANGLE.

When, as in figure 10-20, the sets $P_i$ do not overlap for an initiator set $P$, then the orbit of the IFS is said to be *tiled* by the iterates of the IFS. When an orbit is tiled then the orbit and the limit set are disjoint. A union of just some of the orbital sets is called a *segment* of the orbital set. It is segments that we most often draw when describing an orbit since they can approximate the orbit in our viewing scale. Figure 10-21 shows examples of two orbital sets, one that tiles the orbit and one that does not.

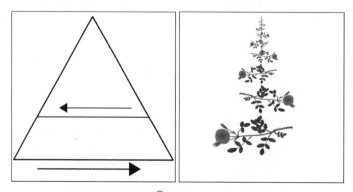

FIGURE 10-20. © ROBERT A. CHAFFER.

FIGURE 10-21. THE TREES IN THE LEFT PANEL GROW ON A FRAMEWORK
OF THE SIERPINSKI TRIANGLE BUT DO NOT TILE THE ORBITAL SET.
IN CONTRAST, THE ORBITAL SET SHOWN ON THE RIGHT IS TILED
BY THE ITERATIONS OF ITS GENERATOR. © ROBERT A. CHAFFER.

## CURVES AND SPACE-FILLING CURVES

Humpty Dumpty, a character in Lewis Carroll's (the pseudonym for the
English mathematician Charles Lutwidge Dodgson [1832–1898]) *Through
the Looking-Glass, and What Alice Found There* (1871), said "When I use
a word, it means just what I choose it to mean—neither more nor less."

Thus it has been with the term *curve*. To the general layman, *curve* is
often taken as a contrast to the words *straight* or *in a line*. In the study of
geometry, the meaning can be quite different and workers in the various
specialties of mathematics often have their own unique definition for this
word. In this chapter, we take *curve* intuitively in the sense of a path that
can be traced. A single curve of this type would not have any breaks but
it might contain sharp corners and it might intersect or retrace itself. This
class of curves would include lines, line segments, arcs of circles, and segments of polygons. The essential condition for being a curve is that it can
be continuously traced so that each of its points is visited at least once.

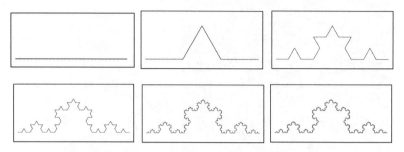

FIGURE 10-22. THE INITIATOR AND THE NEXT FIVE
ORBITAL SETS FOR KOCH'S SNOWFLAKE CURVE.

Mathematicians of the late nineteenth and early twentieth centuries were surprised by the discovery of a number of strange new curves featuring properties that were previously not thought possible for curves. The next several examples show some of these related to triangles and iterated function systems.

As stated above, curves can have sharp corners, but it was once expected that any two such corners of an unbroken curve would have a smooth section between them. Thus, there was considerable interest in a curve announced by the Swedish mathematician Niels Fabian Helge von Koch (1870–1924) in a 1904 paper titled, *On a Continuous Curve without Tangents, Constructible from Elementary Geometry*. His new curve was shown to consist entirely of corner points. Example 12 shows how it is constructed by use of an IFS.

*Example 12.* The Koch snowflake can be described as the limit set of an IFS of four transformations. These are defined in terms of an equilateral triangle identified with four similar sub-triangles as shown in the first panel of figure 10-23. The similarity ratio is $\frac{1}{3}$. When the base of the large triangle is taken as the initiator, the IFS iteration results in continuous curves in the orbit of the IFS (see figure 10-22). The limit set is the Koch curve. It is fractal with self-similarity dimension $\frac{\log 4}{\log 3} = 1.262$. Although the details that show that the limit set is actually a curve (not true for all infinite sequences of curves that converge to a limit set) and that it consists entirely of corners are quite technical, the sequence of approximations does clearly suggest how the many corners are created. It can be seen that the corners introduced do not get deleted in later steps, so the number of corners does increase without bound and they become packed ever closer together.

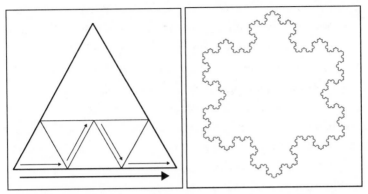

FIGURE 10-23. THE PLAN FOR THE SNOWFLAKE IFS (*LEFT*) AND THREE COPIES OF THE SNOWFLAKE CURVE PLACED AROUND A TRIANGLE FORM KOCH'S SNOWFLAKE (*RIGHT*).

The curve, gamma, in the first panel of figure 10-25 is quite special. Its uniqueness arises from what happens when it is used as an initiator under the IFS, defined by figure 10-24. This curve enters triangle $ABC$ at vertex $A$ and follows a path through the triangle and exits at vertex $B$. Under the IFS each image in the orbit of this initiator is also a curve that begins at $A$ and ends at $B$ while the full limit set is triangle $ABC$.

As a consequence of the above meaning of curve, there will be specific points on the curve that are $\frac{1}{2}$ of the way from $A$ to $B$, $\frac{3}{5}$ the way from $A$ to $B$, etc. Panel 1 of figure 10-25 indicates the halfway point by an arrow and the $\frac{3}{5}$ point with a dot. Each of the next consecutive curves in the orbit of gamma also have $\frac{1}{2}$ and $\frac{3}{5}$ points. In the panels of figure 10-25, the halfway points are indicated by arrows. The $\frac{3}{5}$ points are indicated by dots. For purposes of comparison, the previous location indicators of $\frac{3}{5}$ points are retained in successive panels.

Because later iterations of the curve self-intersect, it becomes more difficult to discern the path of travel along each curve as imposed by the IFS. The dots shown are results of careful analysis of the MRCM to determine the proper locations.

The halfway point makes a large change in location from the initiator to the first iterate. After that, it does not change location as iteration continues. The entry and exit points do not move at all. The behavior of the $\frac{3}{5}$ point is quite different. It continues to change locations within triangle *ABC* throughout the infinitely many iterations of the MRCM. Even so, its behavior is organized. The jumps between its successive locations become ever smaller as iteration progresses. An analysis of this behavior makes it possible to identify a unique location as $\frac{3}{5}$ the way through the curve after infinitely many iterations, in other words, $\frac{3}{5}$ the way through the limit set.

The eventual stabilization of the locations for $\frac{1}{2}$ and $\frac{3}{5}$ points is not exceptional. For each fraction of the distance from *A* to *B* there is a unique point within triangle *ABC* that corresponds to that fraction. Some of these distance markers, like the $\frac{1}{2}$ point, come to a rest after a number of iterations, and others, like the $\frac{3}{5}$ point, continue to wander but move less and less, on each iteration so that a final resting place can be determined. It is this fact that led mathematicians to view the limiting curve as a special sort, a space-filling curve.

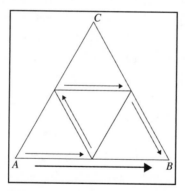

FIGURE 10-24. A PLAN FOR AN IFS THAT FILLS TRIANGLE *ABC*,
SIMILAR BUT DIFFERENT FROM THE IFS IN EXAMPLE 7.

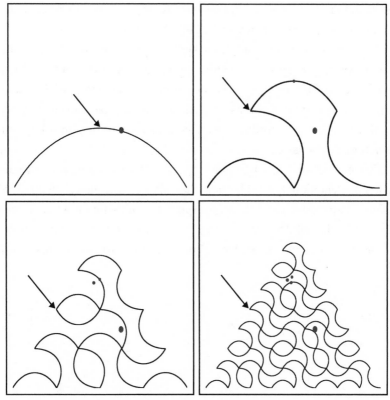

FIGURE 10-25. *TOP LEFT TO BOTTOM RIGHT*, $\Gamma$, $\Gamma_1$, $\Gamma_2$, $\Gamma_3$, SHOWING LOCATION OF $\frac{1}{2}$ AT EACH ITERATION (ARROW) AND THE SUCCESSIVE LOCATIONS OF $\frac{3}{5}$ (DOTS) AS THE $\frac{3}{5}$ LOCATION CONVERGES TOWARD ITS FINAL POSITION.

The existence of such curves came as quite a surprise to the mathematical community when first discovered in 1890 by the Italian mathematician Giuseppe Peano (1858–1932). The expectation of the time was that an essentially one-dimensional object (a curve) could not occupy all of a two-dimensional object that has measurable area. It is now known that an IFS can be used to construct many (but not all) such examples. The original methods were somewhat different. After the discovery of the first example, additional space-filling curves were announced, many having additional surprising properties. A detailed account of these developments is chronicled in a book by Sagen (Hans Sagen, *Space-Filling Curves* [Berlin, Germany: Springer-Verlag, 1994]).

Because of their intricacy and symmetry, space-filling curves have become popular images for creation with the help of computer software. The example of figure 10-25 has self-similar dimension 2 so that, unlike many examples of such curves, it might not always be considered to be a fractal. Example 13, introduced by Knopp, shows a triangle-filling curve of fractional dimension.

*Example 13.* Figure 10-24 demonstrates another triangle-filling curve. It is a representation by the German mathematician Konrad Knopp (1882–1957) of a space-filling curve of Sierpinski's. In this case, the curves are not completely defined by an IFS. Following each individual iteration, a connecting segment is needed to join the two curve segments that are generated.

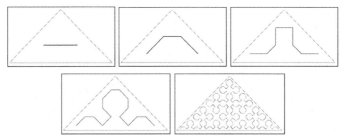

FIGURE 10-26. *TOP LEFT TO BOTTOM RIGHT,* $\Pi$, $\Pi_1$, $\Pi_2$, $\Pi_4$, AND $\Pi_8$.

*Example 14.* Space-filling curves can fill fractal sets just as they can fill traditional geometric figures. Figure 10-27 shows the definition of an IFS, the second and forth iterates of an initiator curve (an arc joining the bottom left to bottom right of the large triangle). The fractal that this curve fills appears to be somewhat similar to the Sierpinski triangle but is still not that triangle. From its resemblance to the standard nuclear energy icon we might label that fractal as the "nuclear triangle." This fractal can be derived from a tiling of the equilateral triangle into nine similar sub-triangles. An appropriate six-transformation IFS, applied in a manner similar to the idea used in example 11, generates this fractal. As in that example, the self-similarity dimension is 1.631. The nuclear fractal also relates to the Pascal triangle. In that discussion, a 0-1 triangle was constructed by replacing multiples of 2 among the Pascal triangle entries with 0s and the other entries with 1s. If, instead, multiples of 3 are replaced by 0s and the others by 1s, then the pattern of the nuclear fractal emerges.

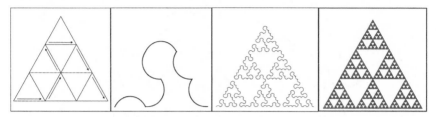

FIGURE 10-27

A final appropriate warning is that it is easy to propose IFS-based examples of curves that converge to fill a portion of space but actually fail to have a true curve as a limit. (For an example, see Heinz-Otto Peitgen, Hartmut Jürgens, and Dietmar Saupe, *Chaos and Fractals: New Frontiers of Science* [New York: Springer-Verlag, 2004], pp. 98–101.)

## THE CHAOS GAME

Generation of a fractal image by use of an MRCM is a computational task of considerable complexity. A surprising alternative is an IFS-based algorithm called the "chaos game" or, alternately, a "fortune-wheel reduction copy machine" or FRCM.

While we most often think of the MRCM from an IFS as acting on an initiator containing a multitude of points, it applies equally well on a set containing only a single point. It has already been pointed out that even a one-element initiator leads to the same limit set when the transformations are contractive. This can work because the number of points increases rapidly as the MRCM iterates. The diagram in figure 10-28 shows how the "children" of a point $P$ in a one-element initiator set, $\Pi$, multiply under the IFS of example 1. This is not a picture of the geometric locations in the triangle $ABC$ but rather an enumeration of the reproduction events. The second row in the tree shown represents the set of points in $\Pi_1$ and the third (shown staggered) row represents the set of points in $\Pi_2$.

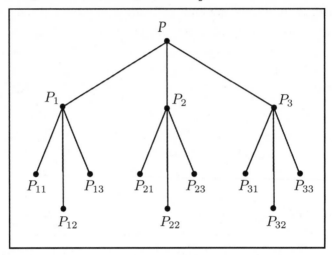

FIGURE 10-28. AN ENUMERATION OF THE CHILDREN OF POINT $P$ IN $\Pi_2$.

The number of points at each level triples since $k = 3$ in this IFS. The sequence of point-population sizes as the iterations proceed is $1, 3, 27, 81,$ $243, \ldots$ or equivalently, $3^0, 3^1, 3^2, 3^3, 3^4, \ldots$. The number of points in $\Pi_{12}$ is over half a million!

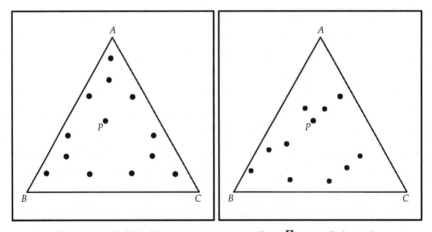

FIGURE 10-29. THE CHILDREN OF $P$ IN $\Pi_1$ AND 2 (LEFT).
TEN GENERATIONS OF $P$ GENERATED DURING A RANDOM TRIP DOWN
THE DECISION TREE (RIGHT).

Figure 10-29 illustrates what results from *P* as the MRCM is applied. The left panel shows all the locations affected by the first three levels of the decision tree, and the panel at the right shows the effect of a random trip down ten levels of the tree to a particular point. At each node in the tree, a random choice from the three transformations has been applied. Not enough points are shown to visually detect much difference, but in the left panel we begin to see a skeleton of the Sierpinski triangle, while in the right panel much less organization is evident. The process illustrated in the right panel is usually referred to as the *chaos game* and is typically described as follows.

> A random starting point is selected. A randomizing device such as a die is employed to select randomly from the set {1,2,3}. Corresponding to this random selection, one of $F_A$, $F_B$, or $F_C$ is applied to the point. In the case of example 1 this could also be specified as "move half the distance from the current location toward vertex *A*, *B*, or *C* according to the random choice." The resulting new point is plotted inside the triangle and the random application of a transformation is applied to this new point. The process should continue for a huge number of repetitions, perhaps hundreds of thousands or even millions.

The surprising outcome of this game is that, in spite of the randomness of the selections and the seemingly chaotic path as we move from point to point, the result is a quite good representation of the Sierpinski triangle. The technique is general. It can be used with any IFS, randomly selecting from its particular set of transformations, and the result will be an approximate picture of the fixed set of that IFS.

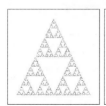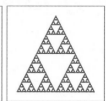

FIGURE 10-30. OUTPUT OF THE CHAOS GAME FOR
10,000, 25,000, 100,000, AND 250,000 ITERATIONS.

The effectiveness of the chaos game for generating the image of a fractal set is initially surprising. Two main ideas explain why it works. The use of a contractive IFS is one of these, and the other depends on probability distribution to complete all regions of the image. (For details see Heinz-Otto Peitgen, Hartmut Jürgens, and Dietmar Saupe, *Chaos and Fractals: New Frontiers of Science* [New York: Springer-Verlag, 2004], pp. 98–101).

The chaos game is highly effective for exploring fractal sets of many kinds. It is relatively easy to write a computer program to implement it. A simple version can be had with just a dozen or so lines of code in a programming language such as Java®, BASIC®, or Processing®, to name a few. A multitude of ready-made and free or inexpensive chaos-game implementations can be found on the Internet. These are suitable for initial exploration, provided the constraints of the software meet the user's needs.

FIGURE 10-31. IMAGE FROM THE IFS IN EXAMPLE 7 OBTAINED BY USING COLORS IN THE CHAOS GAME. © *ROBERT A. CHAFFER.*

Attractions of the chaos-game method include reasonably accurate results even where computer resources are limited in time, storage, and computing muscle. The transformations of the IFS can be any functions that can be represented algebraically in terms of point coordinates. Even better, since it is easy to incorporate a coloring scheme as the points are plotted, this capability allows us to get insight into how the transformations map points, even in cases for which the limit set is a simple triangle or square.

## AN IFS AND TRIANGLE PLAYGROUND

The ideas in this chapter can be combined to equip a virtual playground for exploration of fractal geometry and iterated function systems. The activities can support recreational mathematics, serious mathematical study and research, or even application to the arts. Those enjoying this venue can be involved at many different levels of mathematical sophistication, and they can modify the set of playground equipment to match the resources available. The playground's boundaries are ill defined, so a comprehensive description is not possible. We close with two elementary examples.

One question regards what can be done in the absence of a computer and special software. This is especially an issue for entry-level investigation. A way to explore manually is to create and use IFS graph paper. Figure 10-32 shows one type of such paper. These templates can be manufactured in a variety of ways, including hand drawing on standard square-cell graph paper or with computer drawing software. Once the initial patterns are created, a copy machine suffices to produce additional sheets as needed.

The forms in figure 10-32 are suitable to apply an IFS whose component transformations contract toward the corners or center of the sheet using $\frac{1}{2}$ as a reduction ratio. Each of these transformations may also include a rotation by a multiple of 90° or a reflection about a horizontal, vertical, or diagonal axis before the reduction is applied.

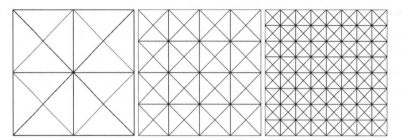

FIGURE 10-32. IFS GRAPH PAPER FOR LEVELS 1, 2, AND 3.

The experiment commences by selecting a number of triangles on the level-1 graph paper and filling them with a suitable color. Based on the transformation recipe of the IFS, this result can then be transformed the required number of times and colored into the appropriate triangles on a level-2 sheet. Proceeding to the following levels in a similar manner often gives a strong indication of the nature of the limit set. Figure 10-33 demonstrates the process for two steps of the IFS in example 10.

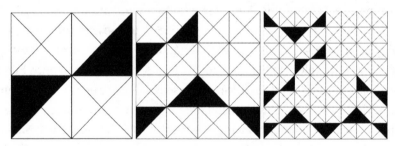

FIGURE 10-33. THE IFS OF EXAMPLE 10 APPLIED (TWO ITERATIONS)
TO AN INITIATOR USING IFS GRAPH PAPER.

The second example is primarily a creative art activity. Copies of a triangle can always be used to repetitively tile the plane. When the triangle is a right, isosceles, or equilateral triangle, then the number of ways of doing the tiling is increased. One or several iterations applied to a chosen initiator using an IFS such as those in examples 9, 10, and 11 produces a base image that can be used to tile the plane in a wallpaper pattern. Figure 10-35 offers a simple example of decorating an IFS iterate to be used in such a tiling. The two wallpaper designs shown in figure 10-34 were created by using an IFS to generate a pattern within a triangle and coloring selected regions from those determined in the triangle. Two copies of the triangle were then merged to form a parallelogram that is laid in rows to tile the plane. Variations in design come primarily from choices of IFS, number of iterations, and coloring patterns in the triangle. Tiling with triangles can also be done in a variety of ways.

FIGURE 10-34. TWO EXAMPLES OF TILING OBTAINED FROM AN IFS AND PARALLELOGRAM TILING OF THE PLANE. © ROBERT A. CHAFFER.

One of the several ways to tile from an equilateral triangle is to rotate the triangle through multiples of 60° in order to achieve a hexagon. Repeating the hexagon tiles the plane. Figure 10-35 shows detail in an example of this idea. The left panel shows a motif obtained by coloring regions bounded by the first iteration of the IFS defined by figure 10-24. Applying the IFS of that figure to the left panel of figure 10-35 produces a basic equilateral triangle that then propagates the plane tiling as shown in the final panel.

FIGURE 10-35. A SELECTION FROM AN IFS ORBIT WITH ARTISTIC
SHADING AND APPLICATION TO TILING THE PLANE WITH HEXAGONS.
© ROBERT A. CHAFFER.

A rich set of explorations awaits those who wish to further investigate the ideas discussed in this chapter. We have used iterated function systems as the unifying theme, but many other options are available. An Internet search on the terms "Pascal triangle," "Sierpinski triangle," "space-filling curves," or "Pascal triangle mod 3" is sure to reward with a wide variety of opportunity.

# APPENDIX

## FOR CHAPTER 2, PAGE 47: TO PROVE THAT AN ANGLE BISECTOR DIVIDES THE OPPOSITE SIDE PROPORTIONALLY TO THE TWO ADJACENT SIDES.

$$\frac{AC}{AB} = \frac{CT_a}{T_aB}$$

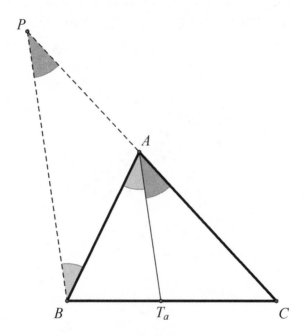

We begin with triangle $ABC$ with angle bisector $AT_a$ (see figure A-1, figure 2-5, and figure 5-3). We construct $BP$ parallel to $AT_a$, meeting $CA$ extended at point $P$. From the parallel lines, we have $\angle CAT_a = \angle APB$, and $\angle T_a AB = \angle ABP$. However, $\angle CAT_a = \angle T_a AB$ $(=\frac{\alpha}{2})$. Therefore, $\angle APB = \angle ABP$, and triangle $ABP$ is isosceles; thus $AB = AP$. From the parallel lines we have $\frac{AC}{AP} = \frac{CT_a}{T_a B}$. Replacing $AB$ for $AP$, we get the sought-after result: $\frac{AC}{AB} = \frac{CT_a}{T_a B}$.

**FOR CHAPTER 3, PAGE 69:**

**NAPOLEON'S THEOREM:**

**IF EQUILATERAL TRIANGLES ARE CONSTRUCTED ON THE SIDES OF ANY TRIANGLE (EITHER OUTWARD OR INWARD) THE CENTERS OF THOSE EQUILATERAL TRIANGLES THEMSELVES FORM AN EQUILATERAL TRIANGLE.**

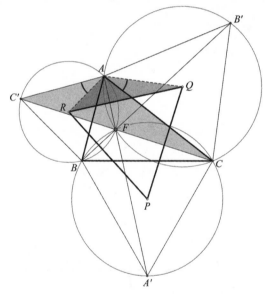

FIGURE A-2

To prove this theorem we consider $\triangle ACB'$ (see figure A-2 and figure 3-5). Since $Q$ is the centroid (point of intersection of the medians) of $\triangle ACB'$, $AQ$ is two-thirds of the length of the altitude (or median). Using the relationships in a $30°$-$60°$-$90°$ triangle, we find that $AC:AQ = \sqrt{3}:1$.

Similarly, in equilateral $\triangle ABC'$, $AC':AR = \sqrt{3}:1$.

Therefore, $AC:AQ = AC':AR$. Also $\angle QAC = \angle RAC' = 30°$, $\angle CAR = \angle CAR$ (reflexive) and therefore, by addition, $\angle QAR = \angle CAC'$.

We can then conclude that $\triangle QAR \sim \triangle CAC'$. It follows that $CC':QR = CA:AQ = \sqrt{3}:1$.

Similarly, we may prove $BB':PQ = \sqrt{3}:1$, and $AA':PR = \sqrt{3}:1$.

Therefore, $BB':PQ = AA':PR = CC':QR$.

But since $BB' = AA' = CC'$ (as proved earlier), we obtain $PQ = PR = QR$.

Thus we can conclude that $\triangle PQR$ is equilateral.

## FOR CHAPTER 5, PAGE 125:
## DERIVATION OF STEWART'S THEOREM.

This theorem yields a relation between the lengths of the sides of the triangle and the length of a cevian of the triangle (see figure 5-19): $a(d^2 + mn) = b^2m + c^2n$.

In $\triangle ABC$, let $BC = a$, $AC = b$, $AB = c$, $CD = d$. Point $D$ divides $BC$ into two segments; $BD = m$ and $CA = n$. Draw altitude $AE = h$ and let $DE = p$.

In order to proceed with the proof of Stewart's theorem, we first derive two necessary formulas. The first one is applicable to triangle $ABD$.

We apply the Pythagorean theorem to triangle $ABE$ to obtain $AB^2 = AE^2 + BE^2$.

Since $BE = m - p$, $c^2 = h^2 + (m - p)^2$.                    (I)

However, by applying the Pythagorean theorem to triangle $ADE$, we have $AD^2 = AE^2 + DE^2$, or $d^2 = h^2 + p^2$, also $h^2 = d^2 - p^2$.

Replacing $h^2$ in equation (I), we obtain

$$c^2 = d^2 - p^2 + (m - p)^2 = d^2 - p^2 + m^2 - 2mp + p^2.$$

Thus, $c^2 = d^2 + m^2 - 2mp$.                    (II)

A similar argument is applicable to triangle $ACD$.

Applying the Pythagorean theorem to triangle $ACE$, we find that $AC^2 = AE^2 + CE^2$.

Since $CE = n + p$, $b^2 = h^2 + (n + p)^2$. $\hspace{2cm}$ (III)

However, $h^2 = d^2 - p^2$, so we substitute for $h^2$ in (III) as follows

$b^2 = h^2 + (n + p)^2 = d^2 - p^2 + n^2 + 2np + p^2 = d^2 + n^2 + 2np$.

Thus, $b^2 = d^2 + n^2 + 2np$. $\hspace{2cm}$ (IV)

Equations (II) and (IV) give us the formulas we need.

Now multiply equation (II) by $n$ to get

$c^2n = d^2n + m^2n - 2mnp$, $\hspace{2cm}$ (V)

and multiply equation (IV) by $m$ to get

$b^2m = d^2m + mn^2 + 2mnp$. $\hspace{2cm}$ (VI)

Adding (V) and (VI), we have

$b^2m + c^2n = d^2m + d^2n + m^2n + mn^2 + 2mnp - 2mnp$,

therefore $b^2m + c^2n = d^2(m + n) + mn(m + n)$.

Since $m + n = a$, we have $b^2m + c^2n = d^2a + mna = a(d^2 + mn)$, which is the relationship we set out to develop.

## FOR CHAPTER 5, PAGE 131:
## PROOF FOR THE SUM OF THE DISTANCES FROM ANY POINT IN A TRIANGLE IN TERMS OF THE SIDE LENGTHS.

$AP^2 + BP^2 + CP^2 = AG^2 + BG^2 + CG^2 + 3GP^2$

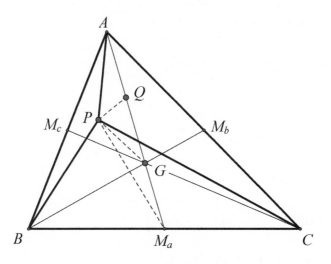

Begin by letting $Q$ be the midpoint of $AG$ (see figure A-3). We now apply the earlier-developed relationship about the medians to each of the following triangles (see chapter 5, p. 129).

$$\Delta PBC: \quad 2PM_a^2 = BP^2 + CP^2 - \frac{BC^2}{2} \tag{I}$$

$$\Delta PAG: \quad 2PQ^2 = AP^2 + GP^2 - \frac{AG^2}{2} \tag{II}$$

$$\Delta PQM_a: \quad 2GP^2 = PQ^2 + PM_a^2 - \frac{QM_a^2}{2} \tag{III}$$

Since $QM_a = \frac{2}{3}AM_a$, and $AG = \frac{2}{3}AM_a$, therefore, $QM_a = AG$.

Substituting in equation (III) and multiplying by 2 we get

$$4GP^2 = 2PQ^2 + 2PM_a^2 - AG^2. \tag{IV}$$

Now adding (I), (II), and (IV):

$$2PM_a^2 + 2PQ^2 + 4GP^2$$
$$= BP^2 + CP^2 - \frac{BC^2}{2} + AP^2 + GP^2 - \frac{AG^2}{2} + 2PQ^2 + 2PM_a^2 - AG^2, \text{ or}$$

$$4GP^2 = BP^2 + CP^2 - \frac{BC^2}{2} + AP^2 + GP^2 - \frac{AG^2}{2} - AG^2, \text{ or}$$

$$AP^2 + BP^2 + CP^2 - 3GP^2 = \frac{3}{2}AG^2 + \frac{1}{2}BC^2. \tag{V}$$

A similar argument made for median $BM_b$ yields

$$AP^2 + BP^2 + CP^2 - 3GP^2 = \frac{3}{2}BG^2 + \frac{1}{2}AC^2. \tag{VI}$$

For median $CM_c$ we get

$$AP^2 + BP^2 + CP^2 - 3GP^2 = \frac{3}{2}CG^2 + \frac{1}{2}AB^2. \tag{VII}$$

By adding (V), (VI), and (VII):

$$3 \cdot (AP^2 + BP^2 + CP^2 - 3GP^2) = \frac{3}{2} \cdot \left(AG^2 + BG^2 + CG^2\right) +$$
$$\frac{1}{2} \cdot \left(BC^2 + AC^2 + AB^2\right). \tag{VIII}$$

We now apply an earlier-developed relationship to triangle $ABC$ (see chapter 5, p. 130), that

$$AM_a^2 + BM_b^2 + CM_c^2 = m_a^2 + m_b^2 + m_c^2 = \frac{3}{4} \cdot \left(a^2 + b^2 + c^2\right)$$

$$= \frac{3}{4} \cdot \left(BC^2 + AC^2 + AB^2\right).$$

Because of $AG = \frac{2}{3}AM_a$, $BG = \frac{2}{3}BM_b$, $CG = \frac{2}{3}CM_c$, we get

$$\frac{9}{4} \cdot \left(AG^2 + BG^2 + CG^2\right) = \frac{3}{4} \cdot \left(BC^2 + AC^2 + AB^2\right), \text{ or}$$

$$3 \cdot \left(AG^2 + BG^2 + CG^2\right) = AB^2 + AC^2 + BC^2.$$

Now substitute this into equation (VIII) to get our desired result:

$$3 \cdot (AP^2 + BP^2 + CP^2 - 3GP^2) = \frac{3}{2} \cdot \left(AG^2 + BG^2 + CG^2\right) +$$
$$\frac{1}{2} \cdot \left[3 \cdot \left(AG^2 + BG^2 + CG^2\right)\right], \text{ also}$$

$$3 \cdot (AP^2 + BP^2 + CP^2 - 3GP^2) = 2 \cdot \frac{3}{2} \cdot \left(AG^2 + BG^2 + CG^2\right)$$
$$= 3 \cdot (AG^2 + BG^2 + CG^2), \text{ or}$$

$$AP^2 + BP^2 + CP^2 = AG^2 + BG^2 + CG^2 + 3GP^2.$$

## FOR CHAPTER 5, PAGE 133:
## THE CENTROID AS BALANCING POINT.
$AX = BY + CZ$

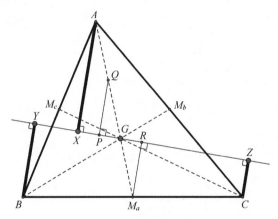

FIGURE A-4

Draw medians $AM_a$, $BM_b$, and $CM_c$ (see figure A-4). From $Q$, the midpoint of $AG$, draw $PQ \perp YZ$. Also draw $RM_a \perp YZ$. Since $\angle AGX = \angle RGM_a$, and $AQ = GQ = GM_a$ (property of a centroid). $\triangle GRM_a \cong \triangle GPQ$, also $RM_a = PQ$. $AX \| CZ$, therefore, $RM_a$ is the median of trapezoid $BYZC$, and $RM_a = \frac{1}{2}(BY + CZ)$ (property of median of a trapezoid). $PQ = \frac{1}{2}AX$ (property of a midline). Therefore, $RM_a = PQ$, $\frac{1}{2}AX = \frac{1}{2}(BY + CZ)$ (transitivity), and $AX = BY + CZ$.

## FOR CHAPTER 6, PAGE 136:
## PROOF OF MENELAUS'S THEOREM
## $AZ \cdot BX \cdot CY = AY \cdot BZ \cdot CX$, IF AND ONLY IF,
## $X$, $Y$, AND $Z$ ARE COLLINEAR.

To prove that if $X$, $Y$, and $Z$ are collinear, then $AZ \cdot BX \cdot CY = AY \cdot BZ \cdot CX$.

Draw a line containing $C$, parallel to $AB$, and intersecting $XYZ$ or $YXZ$ at $D$ (see figures A-5a and A-5b). We are thus beginning with the given collinear points $X$, $Y$, and $Z$.

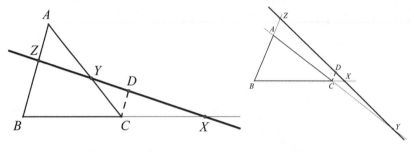

FIGURE A-5a          FIGURE A-5b

For $\triangle CDX \sim \triangle BXZ$, therefore $\dfrac{CD}{BZ} = \dfrac{CX}{BX}$, or $CD = \dfrac{BZ \cdot CX}{BX}$          (I)

For $\triangle CDY \sim \triangle APZ$, therefore $\dfrac{CD}{AZ} = \dfrac{CP}{AY}$, or $CD = \dfrac{AZ \cdot CP}{AY}$          (II)

From equations (I) and (II), $\dfrac{BZ \cdot CX}{BX} = \dfrac{AZ \cdot CP}{AY}$ , from which we easily get $AZ \cdot BX \cdot CY = AY \cdot BZ \cdot CX$.

Now we shall prove that if the points $X$, $Y$, and $Z$ are so situated (with one of the points on the extension of the sides of the triangle) that the equation $AZ \cdot BX \cdot CY = AY \cdot BZ \cdot CX$ is true (or another way of expressing this is that $\frac{AY}{CY} \cdot \frac{BZ}{AZ} \cdot \frac{CX}{BX} = 1$.), then the three points $X$, $Y$, and $Z$ are collinear.

We will let the intersection point of $AB$ and $XY$ be the point $Z'$. Then we have to prove $Z' = Z$.

Because of part 1 (above) we have $\frac{AY}{CY} \cdot \frac{BZ'}{AZ'} \cdot \frac{CX}{BX} = 1$, also $\frac{BZ'}{AZ'} = \frac{BZ}{AZ}$. Therefore is $Z' = Z$, and the points $X$, $Y$, and $Z$ must be collinear.

**FOR CHAPTER 6, PAGE 137:**
**PROOF OF SIMSON'S THEOREM**
**(USING MENALAUS'S THEOREM)**
**THE FEET OF THE PERPENDICULARS DRAWN FROM**
*ANY* **POINT ON THE CIRCUMSCRIBED**
**CIRCLE OF A TRIANGLE TO THE SIDES OF**
**THE TRIANGLE ARE COLLINEAR.**

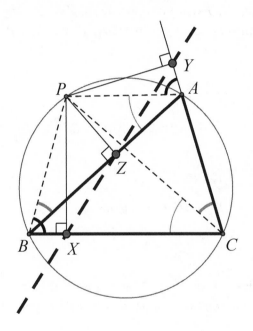

FIGURE A-6

We begin by drawing $PA$, $PB$, and $PC$ (see figure A-6).

$\angle PBA = \frac{1}{2}\overset{\frown}{AP}$, $\angle PCA = \frac{1}{2}\overset{\frown}{AP}$. Therefore $\angle PBA = \angle PCA = \alpha$.

Thus, $\frac{BZ}{PZ} = \cot \alpha = \frac{CY}{PY}$ (in $\triangle BPZ$ and $\triangle CPY$), or $\frac{BZ}{PZ} = \frac{CY}{PY}$,

which implies $\frac{BZ}{CY} = \frac{PZ}{PY}$. 　　　　　　　　　　　　　　　　(I)

Similarly, $\angle PAB = \angle PCB = \beta$ (both are $\frac{1}{2}\overparen{BP}$).

Therefore, $\dfrac{AZ}{PZ} = \cot\beta = \dfrac{CX}{PX}$ (in $\triangle APZ$ and $\triangle CPX$), or $\dfrac{AZ}{PZ} = \dfrac{CX}{PX}$,

which implies $\dfrac{CX}{AZ} = \dfrac{PX}{PZ}$. \hfill (II)

Since $\angle PBC$ and $\angle PAC$ are opposite angles of an inscribed (cyclic) quadrilateral, they are supplementary. However, $\angle PAY$ is also supplementary to $\angle PAC$.

Therefore $\angle PBC = \angle PAY = \gamma$.

Thus, $\dfrac{BX}{PX} = \cot\gamma = \dfrac{AY}{PY}$ (in $\triangle BPX$ and $\triangle APY$), or $\dfrac{BX}{PX} = \dfrac{AY}{PY}$,

which implies $\dfrac{AY}{BX} = \dfrac{PY}{PX}$. \hfill (III)

By multiplying (I), (II), and (III) we obtain

$$\frac{BZ}{CY} \cdot \frac{CX}{AZ} \cdot \frac{AY}{BX} = \frac{PZ}{PY} \cdot \frac{PX}{PZ} \cdot \frac{PY}{PX} = 1.$$

Thus by Menelaus's theorem, $X$, $Y$, and $Z$ are collinear. These three points determine the Simson line of triangle $ABC$ with respect to point $P$.

**FOR CHAPTER 6, PAGE 152:**
**THE CENTER OF THE NINE-POINT CIRCLE OF A TRIANGLE**
**IS THE MIDPOINT OF THE SEGMENT FROM THE**
**ORTHOCENTER TO THE CENTER OF THE CIRCUMCIRCLE.**

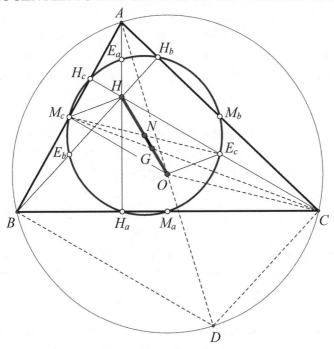

FIGURE A-7

Besides the orthocenter $H$, the centroid $G$, and the circumcenter $O$, we have the nine points on the nine-point circle with center $N$ (see figure A-7):

$M_a, M_b, M_c$ — midpoints of the sides of triangle $ABC$,

$H_a, H_b, H_c$ — feet of the altitudes of triangle $ABC$,

$E_a, E_b, E_c$ — Euler points of triangle $ABC$ (that are the midpoints of the segments between the orthocenter $H$ and the vertices of triangle $ABC$).

Since $E_cM_c$ subtends a right angle at point $H_c$, it must be the diameter of the nine-point circle. Therefore, the midpoint, $N$, of $E_cM_c$ is the center of the nine-point circle.

Extend $AO$ through $O$ to intersect circumcircle $O$ at point $D$. Then draw $BD$ and $CD$. $OM_c$ is a midline of triangle $ABD$. Therefore, $OM_c \| BD$. Since $\angle ABD$ is inscribed in a semicircle, it is a right angle. Now both $BD$ and $CH_c$ are perpendicular to $AB$, so that $BD \| CH_c$. Similarly, $CD \| BH_b$.

We therefore have parallelogram $CDBH$, so that $BD = CH$. Also, $OM_c = \frac{1}{2}BD$ ($OM_c$ is a midline of triangle $ABD$).

Therefore, $OM_c = \frac{1}{2}CH = E_cH$, and $OM_c HE_c$ is a parallelogram (one pair of sides both congruent and parallel). Since the diagonals of a parallelogram bisect each other, the midpoint, $N$, of $E_c M_c$ is also the midpoint of $OH$.

## THE LENGTH OF THE RADIUS OF THE NINE-POINT CIRCLE OF A TRIANGLE IS ONE-HALF THE LENGTH OF THE RADIUS OF THE CIRCUMCIRCLE.

In figure A-7, we notice that $E_c N$ is a midline of triangle $OHC$. Therefore, $E_c N = \frac{1}{2}OC$, which justifies the above statement.

In a paper published in 1765, Leonhard Euler proved that the centroid, $G$, of a triangle trisects the segment $OH$, that is, $OG = \frac{1}{3}OH$. This line, $OH$, is the Euler line of a triangle. (See figures 6-8 and 6-9.)

## THE CENTROID OF A TRIANGLE TRISECTS THE SEGMENT FROM THE ORTHOCENTER TO THE CIRCUMCENTER.

We have already proved that $OM_c \| CH$ (see figure A-7), and we have proved that $OM_c = \frac{1}{2}CH$.

We then have $\triangle OGM_c \sim \triangle HGC$ (AA) with a ratio of similitude of $\frac{1}{2}$. Therefore $OG = \frac{1}{2}GH$, which may be stated as $OG = \frac{1}{3}OH$.

It now remains for us to prove that $G$ is the centroid of triangle $ABC$. From the triangles we just proved similar, we have $GM_c = \frac{1}{2}GC = \frac{1}{3}CM_c$.

But since $CM_c$ is a median, $G$ must be the centroid, because it appropriately trisects the median.

[An alternative proof is very short and elegant, when we use the dilation $D(G, -\frac{1}{2})$ with center $G$ and dilation factor $-\frac{1}{2}$. Here triangle $M_a M_b M_c$ is the image of triangle $ABC$. Therefore the altitudes of triangle $M_a M_b M_c$

are the images of the altitudes of triangle *ABC*. That means the image of the orthocenter *H* is the intersection point *O* of the perpendiculars; also *H*, *G*, and *O* are collinear, and we have *HG* = 2 *GO*.]

It is interesting to note that $\dfrac{HN}{NG} = \dfrac{3}{1} = \dfrac{HO}{OG}$.

We can then see that *HG* is divided internally by *N* and externally by *O* in the same ratio. This is known as a *harmonic* division.

## ALL TRIANGLES INSCRIBED IN A GIVEN CIRCLE AND HAVING A COMMON ORTHOCENTER ALSO HAVE THE SAME NINE-POINT CIRCLE.

Because all triangles inscribed in a given circle and having a common orthocenter also must have the same Euler line, the center of the nine-point circle for all these triangles is fixed at the midpoint of the segment *OH* on the Euler line (established above). Since the radius of the nine-point circle for each of these triangles is half the length of the circumradius (see above), they all have their nine-point circle with the same radius as well as a fixed center. Thus they all must have the same nine-point circle.

## TANGENTS TO THE NINE-POINT CIRCLE OF A TRIANGLE AT THE MIDPOINTS OF THE SIDES OF THE TRIANGLE ARE PARALLEL TO THE SIDES OF THE ORTHIC TRIANGLE.

(See p. 154.)

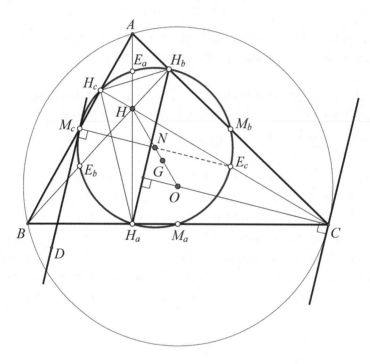

Radius $NM_c$ of the nine-point circle is perpendicular to tangent $DM_c$ (see figure A-8). The circumradii, which contain the vertices of the triangle $ABC$, are perpendicular to the corresponding sides of the orthic triangle $H_aH_bH_c$. Therefore, $OC \perp H_aH_b$.

We showed earlier that $E_cN$ is a midline of triangle $OHC$, and therefore, $E_cN \parallel OC$.

This implies that $E_cNM_c \parallel OC$. Thus $DM_c \parallel H_aH_b$. The proof for the remaining two sides of the orthic triangles is done in the same manner as that shown above.

## TANGENTS TO THE NINE-POINT CIRCLE AT THE MIDPOINTS OF THE SIDES OF THE GIVEN TRIANGLE ARE PARALLEL TO THE TANGENTS TO THE CIRCUMCIRCLE AT THE OPPOSITE VERTICES OF THE GIVEN TRIANGLE.

Since the tangents to the circumcircle at a vertex of the triangle and the tangents to the nine-point circle at the midpoints of the sides of the triangle are each parallel to the sides of the orthic triangle, they are parallel to each other.

An *orthocentric system* consists of four points, each of which is the orthocenter of the triangle formed by the remaining three. In figure A-8, the points $A, B, C$, and $H$ form an orthocentric system, since

$H$   is the orthocenter of triangle $ABC$,
$A$   is the orthocenter of triangle $BCH$,
$B$   is the orthocenter of triangle $ACH$, and
$C$   is the orthocenter of triangle $ABH$.

(See p. 154.)

## THE FOUR TRIANGLES OF AN ORTHOCENTRIC SYSTEM HAVE THE SAME NINE-POINT CIRCLE.

The proof of this property is left to the reader, since all that is required is to check to see if, for each of the four triangles, the nine determining points all lie on the same circle $N$ (see figure A-8, and see figures 6-20 and 6-21).

## THE NINE-POINT CIRCLE OF A TRIANGLE IS TANGENT TO THE INCIRCLE AND THE EXCIRCLES OF THE TRIANGLE.

$N$ *is* the center of the nine-point circle, $I$ the center of the inscribed circle, $H$ the orthocenter, and $O$ the center of the circumscribed circle (see figure A-9, and see figures 6-11 and 6-22; see p. 156):

$F_i$ — point of tangency with the inscribed circle (at $I$),
$F_a$ — point of tangency with the escribed circle (at side $a$),
$F_b$ — point of tangency with the escribed circle (at side $b$),
$F_c$ — point of tangency with the escribed circle (at side $c$).

The proof of this property is quite complex and time consuming. The interested reader will find four different proofs of Feuerbach's theorem in *Modern Geometry*, by Roger A. Johnson ([Boston, MA: Houghton Mifflin, 1929], pp. 200–205).

The proof that Feuerbach actually used consists of computing the distances between the center of the nine-point circle and the centers of the inscribed circle ($r$), the circumscribed circle ($R$), and the circle inscribed in the orthic (or pedal) triangle, $H_a H_b H_c$ (radius $p$), and showing that they equal the sum and difference of the corresponding radii.

$$OI^2 = R^2 - 2Rr \ (Euler)$$

$$IH^2 = 2r^2 - 2Rp$$

$$OH^2 = R^2 - 4Rp$$

$$IN^2 = \frac{1}{2}\left[OI^2 + HI^2\right] - \left(NH\right)^2, \text{ or}$$

$$IN^2 = \frac{1}{4}R^2 - Rr + r^2 = \left(\frac{R}{2} - r\right)^2$$

[N.B. $I$ is the center of the inscribed circle, $G$ is the centroid, $H$ is the orthocenter, and $O$ is the center of the circumscribed circle.]

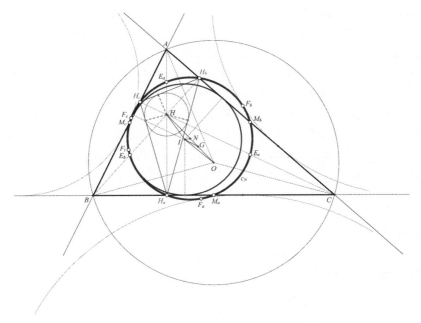

FIGURE A-9

## FOR CHAPTER 6, PAGE 158:
## PROOFS OF MORLEY'S THEOREM:
## IN ANY TRIANGLE, THE THREE POINTS OF
## INTERSECTION OF THE ADJACENT ANGLE TRISECTORS
## FORM AN EQUILATERAL TRIANGLE.

(See chapter 6, figure 6-24.)

We present three proofs.

### PROOF 1:

(Y. Hashimoto, "A Short Proof of Morley's Theorem," *Elemente der Mathematik* 62 [2007]: 121.)

Hashimoto's original proof is short and without a figure:

Let $\alpha$, $\beta$, and $\gamma$ be arbitrary positive angles with $\alpha + \beta + \gamma = 60°$. For any angle $\eta$ we put $\eta' \coloneqq \eta + 60°$. Let triangle *DEF* be an equilateral triangle, and $A$ [resp. $B$, $C$] be the point lying opposite to $D$ [resp. $E$, $F$] with respect to $EF$ [resp. $FD$, $DE$] and satisfying $\angle AFE = \beta'$, $\angle AEF = \gamma'$ [resp. $\angle BDF = \gamma'$, $\angle BFD = \alpha'$; $\angle CED = \alpha'$, $\angle CDE = \beta'$]. Then $\angle EAF = 180° - (\beta' + \gamma') = \alpha$, and similarly $\angle FBD = \beta$, $\angle DCE = \gamma$. By symmetry it is enough to show that $\angle BAF = \alpha$ and $\angle ABF = \beta$ as well.

The perpendiculars from $F$ to $AE$ and $BD$ have the same length $s$. If the perpendicular from $F$ to $AB$ has length $h < s$, then $\angle BAF < \alpha$ and $\angle ABF < \beta$. If, on the other hand, $h > s$, then $\angle BAF > \alpha$ and $\angle ABF > \beta$. Since $\angle BAF + \angle ABF = \alpha' + \beta' + 60° - 180° = \alpha + \beta$, we see that necessarily $h = s$ and $\angle BAF = \alpha$, $\angle ABF = \beta$.

## PROOF 2:

Hashimoto's proof with a figure:

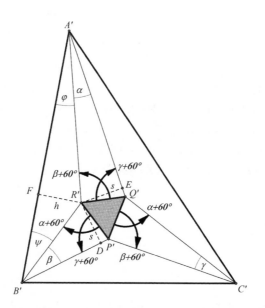

**FIGURE A-10**

Let $\alpha$, $\beta$, and $\gamma$ be random angle measures, with $\alpha + \beta + \gamma = 60°$.

For each of these angles we set the following: $\alpha' := \alpha + 60°$, $\beta' := \beta + 60°$, $\gamma' := \gamma + 60°$. We have the triangle $P'Q'R'$ as equilateral. On each of the sides of the triangle $P'Q'R'$ we draw the triangles $Q'A'R'$, $R'B'P'$, and $P'C'Q'$ so that the following angles are produced:

$$\angle Q'A'R' = \alpha,\ \angle Q'R'A' = \beta' = \beta + 60°,\ \angle A'Q'R' = \gamma' = \gamma + 60°,$$
$$\angle P'B'R' = \beta,\ \angle R'P'B' = \gamma' = \gamma + 60°,\ \angle B'R'P' = \alpha' = \alpha + 60°,$$
$$\angle Q'C'P' = \gamma,\ \angle P'Q'C' = \alpha' = \alpha + 60°,\ \angle Q'P'C' = \beta' = \beta + 60°.$$

It then follows that $\angle Q'A'R' = 180° - (\beta' + \gamma') = 180° - (\beta + 60° + \gamma + 60°)$ $= 60° - (\beta + \gamma) = \alpha$, and analogously, $\angle P'B'R' = \beta$, $\angle Q'C'P' = \gamma$. As a result of the symmetry it suffices to show that $\varphi = \angle R'A'B' = \alpha$ and $\psi = \angle R'B'A' = \beta$.

The perpendicular from $R'$ to $A'Q'$ and $B'P'$ have the same length, $s$. If the perpendicular distance from $R'$ to $A'B'$ is $h$, where $h < s$, then the following is true: $\varphi = \angle R'A'B' < \alpha$ and $\psi = \angle R'B'A' < \beta$.

On the other hand if $h > s$ then it follows that

$$\varphi = \angle R'A'B' > \alpha \text{ and } \psi = \angle R'B'A' > \beta.$$

However, then $\angle A'R'B' = 360° - \angle A'R'Q' - \angle Q'R'P' - \angle P'R'B'$ $= 360° - (\beta + 60°) - 60° - (\alpha + 60°) = 180° - (\alpha + \beta)$.

Thus, $\varphi + \psi = 180° - \angle A'R'B' = 180° - 180° - (\alpha + \beta) = \alpha + \beta$.

Therefore it follows that $h = s$ and $\varphi = \angle R'A'B' = \alpha$, $\psi = \angle R'B'A' = \beta$.

## PROOF 3:

(L. Bankoff, "A Simple Proof of the Morley Theorem," *Mathematics Magazine* 35, no. 4 [1962]: 223–24.)

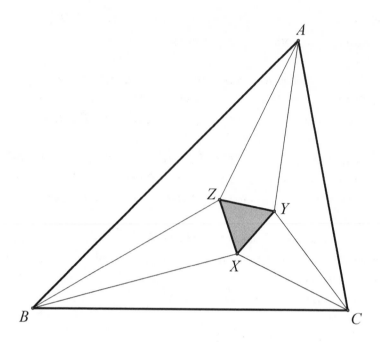

**FIGURE A-11**

In figure A-10, $\sin \angle AYC = \sin\left(\pi - \dfrac{\alpha+\gamma}{3}\right) = \sin\left(\dfrac{\alpha+\gamma}{3}\right) = \sin\left(\dfrac{\pi-\beta}{3}\right) = \sin\left(\dfrac{2\pi+\beta}{3}\right).$

$$\sin 3\theta = \sin\theta \cos 2\theta + \sin 2\theta \cos\theta = \sin\theta\,(3\cos^2\theta - \sin^2\theta)$$

$$= 4\sin\theta\left(\dfrac{\sqrt{3}\cos\theta + \sin\theta}{2}\right)\left(\dfrac{\sqrt{3}\cos\theta - \sin\theta}{2}\right) = 4\sin\theta \cdot \sin\left(\dfrac{\pi}{3}+\theta\right) \cdot \sin\left(\dfrac{\pi}{3}-\theta\right).(*)$$

From the law of sines for triangle $AYC$,

$$AY \cdot \sin\left(\dfrac{\pi-\beta}{3}\right) = AC \cdot \sin\left(\dfrac{\gamma}{3}\right) = 2R \cdot \sin\beta \cdot \sin\left(\dfrac{\gamma}{3}\right),$$ where $R$ is the circum-radius of triangle $ABC$.

Therefore, by (*) $AY = 8R \cdot \sin\left(\dfrac{\beta}{3}\right) \cdot \sin\left(\dfrac{\gamma}{3}\right) \cdot \sin\left(\dfrac{\pi+\beta}{3}\right)$. Similarly,

$$AZ = 8R \cdot \sin\left(\dfrac{\gamma}{3}\right) \cdot \sin\left(\dfrac{\beta}{3}\right) \cdot \sin\left(\dfrac{\pi+\gamma}{3}\right).$$

Therefore, $\dfrac{AZ}{AY} = \dfrac{\sin\left(\dfrac{\pi+\gamma}{3}\right)}{\sin\left(\dfrac{\pi+\beta}{3}\right)}$ .

But $\angle AZY + \angle AYZ = \pi - \dfrac{\alpha}{3} = \dfrac{2\pi+\beta+\gamma}{3} = \dfrac{\pi+\beta}{3} + \dfrac{\pi+\gamma}{3}$ .

From here, $\angle AZY = \dfrac{\pi+\beta}{3}$ and $\angle AYZ = \dfrac{\pi+\gamma}{3}$ , and similarly for triangles $BXZ$ and $CXY$. It thus follows that the sum of angles around $X$, excluding $\angle YXZ$ is 300°, or $\angle YXZ = 60°$. The other two angles are similarly shown to be 60°.

## FOR CHAPTER 7, PAGE 168: [BEFORE FIGURE 7-9] DERIVATION OF HERON'S FORMULA FOR THE AREA OF A TRIANGLE.

$$Area_{\triangle ABC} = \sqrt{s(s-a)(s-b)(s-c)}\text{ , where }s = \dfrac{a+b+c}{2}$$

There is nice geometric proof—actually attributed to Heron—provided in Thomas Heath's *A Manual of Greek Mathematics* (New York: Dover, 1963). However, to conserve space we will provide a much shorter trigonometric derivation building on the two high-school-level relationships: The law of cosines, $c^2 = a^2 + b^2 - 2ab\cos C$; and the Pythagorean identity, $\sin^2\theta + \cos^2\theta = 1$. We had just previously developed the area of a triangle formula as $Area\ \triangle ABC = \frac{1}{2}ab\sin C$.

With $\cos^2 C = \dfrac{(a^2+b^2-c^2)^2}{4a^2b^2}$ and substituting for $\sin C$ in this last equation for the area, we get

$$Area\ \triangle ABC = \tfrac{1}{2}ab\sin C = \tfrac{1}{2}ab\sqrt{1-\cos^2 C} = \tfrac{1}{2}ab\sqrt{1-\dfrac{(a^2+b^2-c^2)^2}{4a^2b^2}}$$

$$= \tfrac{1}{2}ab\sqrt{\dfrac{4a^2b^2-(a^2+b^2-c^2)^2}{4a^2b^2}} = \tfrac{1}{4}\sqrt{4a^2b^2-(a^2+b^2-c^2)^2}\ .$$

We now have to factor the term under the radical sign:

$$4a^2b^2 - (a^2 + b^2 - c^2)^2 = -(a + b + c)\cdot(a + b - c)\cdot(a - b - c)\cdot(a - b + c)$$

$$= -(a + b + c)\cdot(a + b - c)\cdot[-(-a + b + c)]\cdot(a - b + c) = (a + b + c)\cdot$$
$$(a + b - c)\cdot(-a + b + c)\cdot(a - b + c).$$

With $a + b + c = 2s$, $a + b - c = 2(s - c)$, $-a + b + c = 2(s - a)$, and $a - b + c = 2(s - b)$ we get

$$Area\ \triangle ABC = \frac{1}{4}\sqrt{2s\cdot 2(s-c)\cdot 2(s-a)\cdot 2(s-b)} = \sqrt{s\cdot(s-a)\cdot(s-b)\cdot(s-c)}.$$

## FOR CHAPTER 9, PAGE 278:
## ERDÖS-MORDELL INEQUALITY
$PA + PB + PC \geq 2(PD + PE + PF)$

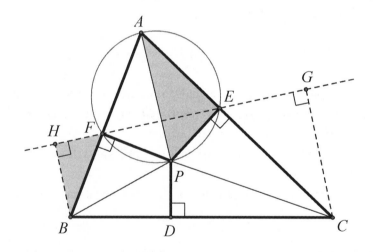

**FIGURE A-12**

Our proof is based on one by H. Lee ("Another Proof of the Erdös-Mordell Theorem," *Forum Geometricorum* 1 [2001]: 7–8), and it applies Ptolemy's theorem (see figure 9-19 and p. 278).

We begin the proof by drawing *HB* and *GC* perpendicular to line *HFEG*. *BC* is either equal to *HG* (if *BHGC* is a rectangle) or is greater than *HG*, if not. Symbolically, $BC \geq HG = HF + FE + EG$.

Since $AFPE$ is a cyclic quadrilateral,[1] $\angle APE + \angle AFE$ (angles at the circumference). However, $\angle AFE = \angle BFH$. Therefore, $\angle APE = \angle BFH$, thereby determining that $\triangle BFH \sim \triangle APE$.

This gives us

$$HF = \frac{PE}{PA} \cdot BF .$$

Using this procedure, we can show that

$$EG = \frac{PF}{PA} \cdot CE .$$

The famous Ptolemy theorem,[2] which for quadrilateral $AFPE$ gives us

$PA \cdot FE = AF \cdot PE + AE \cdot PF$, can be rewritten as $FE = \frac{AF \cdot PE + AE \cdot PF}{PA}$ .

With appropriate substitutions, we get $BC \geq \frac{PE}{PA} \cdot BF + \frac{AF \cdot PE + AE \cdot PF}{PA} + \frac{PF}{PA} \cdot CE$.

This can be rewritten as

$$PA \cdot BC \geq PE \cdot BF + AF \cdot PE + AE \cdot PF + PF \cdot CE$$
$$= PE\,(BF + AF) + PF\,(AE + CE) = PE \cdot AB + PF \cdot AC.$$

Dividing by $BC$, we have $PA \geq PE \cdot \dfrac{AB}{BC} + PF \cdot \dfrac{AC}{BC}$ .

If we were to use the other two sides of triangle $ABC$ from which to draw the perpendiculars as we did for $BC$, we would get $PB \geq PF \cdot \dfrac{BC}{CA} + PD \cdot \dfrac{BA}{CA}$, and $PC \geq PD \cdot \dfrac{CA}{AB} + PE$ .

Using the fact that for positive real numbers $m$ and $n$ it follows that $\frac{n}{m} + \frac{m}{n} \geq 2$, we can conclude that $PA + PB + PC \geq 2(PD + PE + PF)$. Of course, if and only if triangle $ABC$ is equilateral and $P$ is the center of the circumscribed circle then $PA + PB + PC = 2(PD + PE + PF)$.

**FOR CHAPTER 9, PAGE 282:**

**TO PROVE THAT** $c \geq \dfrac{1}{\sqrt{2}} \cdot (a+b) = \dfrac{\sqrt{2}}{2}(a+b).$

We begin with right triangle $ABC$, where $\angle C = 90°$, $\angle A = \alpha$, and $\angle B = \beta$ (see figure A-3). Also recall that $AB = c$, $BC = a$, and $AC = b$.

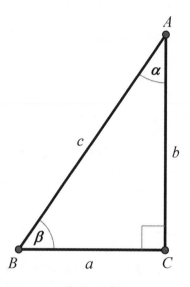

FIGURE A-13

For the proof we will use the trigonometric functions $\cos \alpha = \dfrac{b}{c}$, and $\cos \beta = \dfrac{a}{c}$. It then follows that

$$a + b = c \cdot (\cos \beta + \cos \alpha) = c \cdot 2 \cdot \cos \frac{\alpha+\beta}{2} \cdot \cos \frac{\alpha-\beta}{2}$$
$$= c \cdot 2 \cdot \cos 45° \cdot \cos \frac{\alpha-\beta}{2} = c \cdot 2 \cdot \frac{\sqrt{2}}{2} \cdot \cos \frac{\alpha-\beta}{2}$$

$= c \cdot \sqrt{2} \cdot \cos \frac{\alpha-\beta}{2}$, or, written otherwise, gives us $c = \dfrac{1}{\sqrt{2} \cdot \cos \frac{\alpha-\beta}{2}} \cdot (a+b).$

Since $0 < \cos \frac{\alpha-\beta}{2} \leq 1$, we can state that $c = \dfrac{1}{\sqrt{2} \cdot \cos \frac{\alpha-\beta}{2}} \cdot (a+b) \geq \dfrac{1}{\sqrt{2}} \cdot (a+b).$

The equality only holds true when $a = b$, which would be an isosceles right triangle. Considering that $c < a + b$, we get from the above inequality that $0 < (a + b) - c \leq \left(1 - \dfrac{1}{\sqrt{2}}\right) \cdot (a+b)$, or $0 < (a + b) - c \leq \dfrac{2-\sqrt{2}}{2} \cdot (a+b).$

# NOTES

## CHAPTER 1: INTRODUCTION TO THE TRIANGLE

1. Isidore Lucien Ducasse, *Maldoror (And the Complete Works of the Comte de Lautréamont)*, translated by Alexis Lykiard (Cambridge, MA: Exact Change, 1994). The original source was *Les Chants de Maldoror*, published in 1874 and written by the French poet Lautréamont, also known as Comte de Lautréamont, both of which are pseudonyms for Isidore Lucien Ducasse (1846–1870).

2. For more on the Pythagorean theorem, see *The Pythagorean Theorem: The Story of Its Power and Beauty,* by Alfred S. Posamentier (Amherst, NY: Prometheus Books, 2010).

3. The word *obtuse* used outside of a mathematical context means *dull*. Just as an obtuse angle is a dull angle.

4. The word *acute* used outside of a mathematical context means *sharp*. Just as an acute angle is a sharp angle.

5. You can find a proof at *Wikipedia*, s.v. "Apollonius' theorem," last modified April 11, 2012, http://en.wikipedia.org/wiki/Apollonius%27_theorem. This relationship is a special case of Stewart's theorem (see chap. 5, p. 121).

6. A proof of this formula was first produced in the book *Metrica* by Heron of Alexandria in circa 60 CE. More recently, writings of the Arab scholar Abu'l Raihan Muhammed al-Biruni have credited the formula to Heron's predecessor Archimedes prior to 212 BCE (Eric W. Weisstein, "Heron's Formula," MathWorld, http://mathworld.wolfram.com/HeronsFormula.html).

7. In trigonometric terms, the Pythagorean theorem is $\sin^2\angle A + \sin^2\angle B = 1$, or $\sin^2\alpha + \sin^2\beta = 1$.

8. If $\angle A = 90°$, then $\cos \angle A = 0$, therefore we get $a^2 = b^2 + c^2$ (theorem of Pythagoras).

9. For more about this ubiquitous ratio, see *The Glorious Golden Ratio* by A. S. Posamentier and I. Lehmann (Amherst, NY: Prometheus Books, 2012).

10. With some algebraic manipulations we get the following relationships; first, we express $c$ and $a$ in terms of $b$:

$$c = \phi\, b = \frac{\sqrt{5}+1}{2} \times b, \text{ and } a = \phi c = \phi^2 b = \frac{\sqrt{5}+1}{2}\, b,$$

then $a = \frac{\sqrt{5}-1}{2} b$, which is the same as $\phi^2 b = (\phi + 1)b$.

Furthermore, in terms of $c$, we get $a = \phi c = \frac{\sqrt{5}-1}{2} c$ and $b = \phi^{-1} \times c = \frac{1}{\phi} \times c = \frac{\sqrt{5}-1}{2} \times c$.

Then, in terms of $a$, we get $c = \phi^{-1} \times a = \frac{1}{\phi} \times a = \frac{\sqrt{5}-1}{2} \times a$ and $b = \phi^{-1} \times c = \phi^{-2} \times a = \frac{1}{f^2} \times a = \frac{3-\sqrt{5}}{2} \times a$.

# CHAPTER 2: CONCURRENCIES OF A TRIANGLE

1. An *altitude* of a triangle is the line segment from a vertex perpendicular to the opposite side.

2. An *angle bisector* of a triangle is the line segment from a vertex to the opposite side and bisecting the angle.

3. A *median* of a triangle is the line segment joining a vertex with the midpoint of the opposite side.

4. Wilfried Haag, *Wege zu geometrischen Sätzen* (Stuttgart/Düsseldorf/Leipzig, Germany: Klett, 2003), p. 40.

5. This is a biconditional statement that indicates that if the lines are concurrent then the equation is true, and if the equation is true then the lines are concurrent.

6. Albert Gminder, *Ebene Geometrie* (München and Berlin: Oldenbourg, 1932), p. 421.

7. The same proof holds true for both an acute and an obtuse triangle.

8. Carl Adams, *Die Lehre von den Transversalen in ihrer Anwendung auf die Planimetrie. Eine Erweiterung der euklidischen Geometrie* (Winterthur, Switzerland: Druck und Verlag der Steiner'schen Buchhandlung, 1843).

9. A proof by John Rigby can be found in Ross Honsberger, *Episodes in Nineteenth and Twentieth Century Euclidean Geometry* (Washington, DC: Mathematical Association of America, 1995), pp. 63–64.

# CHAPTER 4: CONCURRENT CIRCLES
# OF A TRIANGLE

1. Auguste Miquel, "Mémoire de Géométrie," *Journal de mathématiques pures et appliquées de Liouville* 1 (1838): 485–87.

2. The first elementary proof of this relationship was by William Clifford (1845–1879), and the first algebraic proof was published in 2002 by Hongbo Li ("Automated Theorem Proving in the Homogeneous Model with Clifford Bracket Algebra," *Applications of Geometric Algebra in Computer Science and Engineering*, edited by L. Dorst et al. (Boston, MA: Birkhauser, 2002), pp. 69–78.

3. R. Johnson, "A Circle Theorem," *American Mathematical Monthly* 23 (1916): 161–62.

# CHAPTER 5: SPECIAL LINES
# OF A TRIANGLE

1. In France and England it is called the *Lemoine point* and in Germany the *Grebe point*.

2. The point *G* is here *not* the centroid of the triangle *ABC*.

3. The center of gravity of a triangle is the point at which a cardboard triangle can be balanced on the point of a pin.

# CHAPTER 6: USEFUL TRIANGLE THEOREMS

1. E. H. Lockwood, *A Book of Curves* (London: Cambridge University Press, 1971), pp. 76–79, refers to Jakob Steiner, "Über eine besondere Curve dritter Classe (und vierten Grades)," *Burchardt's Journal Band LIII*, pp. 231–37 (presented at the Academy of Sciences–Berlin, January 7, 1856). Steiner's paper was published in Jacob Steiner, *Gesammelte Werke, Band II*, edited by K. Weierstrass (Berlin, Germany: G. Reimer, 1882), pp. 639–47.

2. One source for more Simson line relationships can be found in *Challenging Problems in Geometry*, by A. S. Posamentier and C. T. Salkind (New York: Dover, 1988).

3. Leonhard Euler, "Solutio facilis problematum quorundam geometricorum difficillimorum," *Novi Commentarii Academiae Scientiarum Imperialis Petropolitanae* 11 (1767): 103–23. Reprinted in *Opera Omnia* 1, no. 26 (1953): 139–57.

According to the Euler Archive, it was presented to the Petersburg Academy on

December 12, 1763 (Euler Archive, "17. All Publciations," http://www.math.dartmouth .edu/~euler/tour/tour_17.html).

4. Another proof of this collinearity can be found in A. S. Posamentier, *Advanced Euclidean Geometry* (Hoboken, NJ: John Wiley, 2002), pp. 161–63.

5. C. J. Brianchon and J.-V. Poncelet, "Géométrie des courbes. Recherches sur la détermination d'une hyperbole équilatère, au moyen de quatres conditions donnée," *Annales de Mathématiques pures et appliquées* 11 (1820–1821): 205–20.

6. Roger A. Johnson, *Advanced Euclidean Geometry* (Mineola, NY: Dover, 1960), p. 200.

7. Howard Eves, *A Survey of Geometry*, rev. ed. (Boston, MA: Allyn & Bacon, 1972; repr. 1965), p. 133.

8. The truly motivated reader might like to see even more points of significance of a triangle. We refer, then, to the website MathWorld, http://mathworld.wolfram.com/ KimberlingCenter.html.

# CHAPTER 7: AREAS OF AND WITHIN TRIANGLES

1. Jens Carstensen, "Die Seitenhalbierenden—Ein schöner Satz," *Die Wurzel*," July 2004, pp. 160–62.

2. The answer is $Area \; \Delta EFG = \dfrac{8}{27} \; Area \; \Delta ABC$.

3. A proof of this can be found at "Napoleon's Theorem," Mathpages, http://www .mathpages.com/home/kmath270/kmath270.htm.

4. Hugo Steinhaus, *Mathematical Snapshots* (Mineola, NY: Dover, 1999), p. 9.

5. Ingmar Lehmann, "Dreiecke im Dreieck. Vermutungen und Entdeckungen—DGS als Wundertüte," *Werkzeuge im Geometrieunterricht*, edited by Andreas Filler, Mathias Ludwig, and Reinhard Oldenburg (Hildesheim/Berlin, Germany: Franzbecker, 2011), pp. 101–20.

6. Edward Routh (1831–1907) was an English mathematician, noted as the outstanding coach of students preparing for the Mathematical Tripos examination of the University of Cambridge. Edward J. Routh, *A Treatise on Analytical Statics, with Numerous Examples*, vol. 1, 2nd ed. (Cambridge, Cambridge University Press, 1896), p. 82.

7. Lehmann, "Dreiecke im Dreieck," pp. 101–20.

# CHAPTER 8: TRIANGLE CONSTRUCTIONS

1. Various constructions of the regular pentagon can be found in A. S. Posamentier and I. Lehmann, *The Glorious Golden Ratio* (Amherst, NY: Prometheus Books, 2012).

2. Gauss used the following relationship:

$$\cos\left(\frac{360°}{17}\right) = -\frac{1}{16} + \frac{1}{16}\sqrt{17} + \frac{1}{16}\sqrt{34 - 2\sqrt{17}} + \frac{1}{8}\sqrt{17 + 3\sqrt{17} - \sqrt{34 - 2\sqrt{17}} - 2\sqrt{34 + 2\sqrt{17}}}$$

3. T. Kempermann, *Zahlentheoretische Kostproben* (Frankfurt am Main, Germany: Harri Deutsch, 2005), p. 35.

4. $\left(\dfrac{15}{3}\right) = \dfrac{15 \cdot 14 \cdot 13}{1 \cdot 2 \cdot 3} = 455$.

5. J. Boehm, W. Börner, E. Hertel, O. Krötenheerdt, W. Mögling, L. Stammler, *Geometrie, II. Analytische Darstellung der euklidischen Geometrie* (Berlin, Germany: DVW, 1975), pp. 203–205; Walter Gellert, Herbert Kästner, Siegfried Neuber, eds., *Lexikon der Mathematik* (Leipzig, Germany: Bibliographisches Institut, 1977), pp. 105–106.

6. In order to construct the product and the quotient of two given line segments $a$ and $b$ you can set up the following procedure:

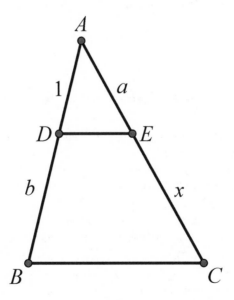

To find the product of the two segments $a$ and $b$, we use $\dfrac{1}{b} = \dfrac{a}{x}$, which gives us $x = ab$.

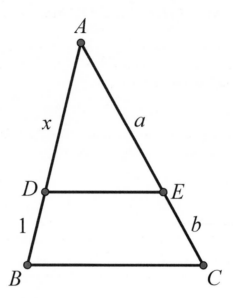

To find the quotient of the two segments $a$ and $b$, we use

$$\frac{x}{1} = \frac{a}{b}, \text{ which is } x = \frac{a}{b}.$$

# CHAPTER 9: INEQUALITIES IN A TRIANGLE

1. $\left(\sqrt{x} - \sqrt{y}\right)^2 \geq 0$ gives us $x + y \geq 2\sqrt{xy}$. With $x + y \geq 2\sqrt{xy}$, $x + z \geq 2\sqrt{xz}$ and $y + z \geq 2\sqrt{yz}$ we get $(y + z)(z + x)(x + y) \geq 2\sqrt{xy} \cdot 2\sqrt{xz} \cdot 2\sqrt{yz} = 8xyz$.

2. This inequality was proposed by Paul Erdös ("Problem 3740," *American Mathematical Monthly* 42 [1935]: 396) and solved two years later by Louis Joel Mordell and D. F. Barrow ("Solution to Problem 3740," *American Mathematical Monthly* 44 [1937]: 252–54).

3. This is known as the theorem of Möbius-Pompeiu, by the German mathematician and astronomer August Ferdinand Möbius (1790–1868) and the Rumanian mathematician Dimitrie Pompeiu (1873–1954).

4. Leonhard Euler (1707–1783); William Chapple (1718–1781).

5. W. J. Blundon, "Problem E1935," *American Mathematical Monthly* 73 (1966): 1122; A. Makowski, "Solution of the Problem E1935," *American Mathematical Monthly* 75 (1968): 404.

6. Developed by the Austrian mathematician Roland Weitzenböck (1885–1955).

7. Developed by the Swiss mathematicians Hugo Hadwiger (1908–1981) and Paul Finsler (1894–1970).

# CHAPTER 10: TRIANGLES AND FRACTALS

1. A. Farina, S. Giompapa, A. Graziano, A. Liburdi, M. Ravanelli, F. Zirilli, "Tartaglia-Pascal's Triangle: A Historical Perspective with Applications," *Signal, Image and Video Processing* (May 24, 2011): 1–16.

2. Kazimir Malevich (1879–1935) was a Russian painter and art theoretician, born of ethnic Polish parents. "He was a pioneer of geometric abstract art and the originator of the Avant-garde Supremacist movement." (*Wikipedia*, s.v. "Kazimir Malevich," last modified May 12, 2012, http://en.wikipedia.org/wiki/Kazimir_Malevich.)

# APPENDIX

1. That is, one that can be inscribed in a circle, since in this case, we have the opposite angles supplementary.

2. Ptolemy's theorem states that for a cyclic quadrilateral the product of the diagonals equals the sum of the products of the opposite sides (see p. 274).

# REFERENCES

## BOOKS

Andreescu, Titu, Oleg Mushkarov, and Luchezar Stoyanov. *Geometric Problems on Maxima and Minima*. Basel, Switzerland: Birkhäuser, 2006.

Boehm, J., W. Börner, E. Hertel, O. Krötenheerdt, W. Mögling, and L. Stammler. *Geometrie*. Vol. 2. *Analytische Darstellung der euklidischen Geometrie*. Berlin, Germany: DVW, 1975.

———. *Aufgabensammlung*. Vol. 2. Berlin, Germany: DVW, 1982.

Coxeter, H. S. M., and S. L. Greitzer. *Geometry Revisited*. Washington, DC: Mathematical Association of America, 1967.

Herterich, K. *Die Konstruktion von Dreiecken*. Stuttgart, Germany: Klett, 1986.

Kazarinoff, Nicholas D. *Geometric Inequalities*. New Haven, CT: Yale, 1961.

Manfrino, Radmila Bulajich, José A. Gómez Ortega, and Rogelio Valdez Delgado. *Inequalities: A Mathematical Olympiad Approach*. Basel, Switzerland: Birkhäuser, 2009.

Mettler, Martin. *Vom Charme der "verblassten" Geometrie*. Timisoara, Romania: Verlag Eurobit, 2000.

Nelsen, Roger B. *Proofs without Words II*. Washington, DC: Mathematical Association of America, 2000.

Posamentier, Alfred S. *Advanced Euclidean Geometry: Excursions for Secondary Teachers and Students*. Emeryville, CA: Key College, 2002.

———. *Making Geometry Come Alive! Student Activities & Teacher Notes*. Thousand Oaks, CA: Corwin Press, 2000.

———. *The Pythagorean Theorem: The Story of Its Power and Beauty*. Amherst, NY: Prometheus Books, 2010.

Posamentier, Alfred S., and C. T. Salkind. *Challenging Problems in Geometry*. New York: Macmillan, 1970. Reprint, Palo Alto, CA: Dale Seymour, 1988; New York: Dover, 1996.

Posamentier, Alfred S., and G. Sheridan. *Math Motivators: Investigations in Geometry*. Menlo Park, CA: Addison-Wesley, 1982.

Posamentier, Alfred S., and Herbert A. Hauptman. *101 Great Ideas for Introducing Key Concepts in Mathematics: A Resource for Secondary School Teachers*. Thousand Oaks, CA: Corwin Press, 2001. Second edition published 2006.

Posamentier, Alfred S., and Ingmar Lehmann. *The Glorious Golden Ratio*. Amherst, NY: Prometheus Books, 2012.

——. *Mathematical Amazements and Surprises: Fascinating Figures and Noteworthy Numbers*. Amherst, NY: Prometheus Books, 2009.

——. π: *A Biography of the World's Most Mysterious Number*. Amherst, NY: Prometheus Books, 2004.

Posamentier, Alfred S., J. H. Banks, and R. L. Bannister. *Geometry, Its Elements and Structure*. New York: McGraw-Hill, 1972. Second edition published 1977.

Posamentier, Alfred S., and W. Wernick. *Advanced Geometric Constructions*. Palo Alto, CA: Dale Seymour, 1988.

Specht, Eckard. *Geometria—Scientiae Atlantis*. Magdeburg, Germany: Otto-von-Guericke-Universität, 2001.

# JOURNALS

*Crux Mathematicorum with Mathematical Mayhem*. Canadian Mathematical Society, Ottawa, Canada.

*Forum Geometricorum: A Journal on Classical Euclidean Geometry and Related Areas*. Department of Mathematical Sciences, Florida Atlantic University.

*Die Wurzel. Zeitschrift für Mathematik*. Department of Mathematics and Computer Science, Friedrich-Schiller University, Jena, Germany.

# INDEX

# READER'S NOTES

# READER'S NOTES

# READER'S NOTES